Embedded C Programming

Embedded C Programming

Techniques and Applications of C and PIC® MCUS

Mark Siegesmund

AMSTERDAM • BOSTON • HEIDELBERG • LONDON • NEW YORK • OXFORD
PARIS • SAN DIEGO • SAN FRANCISCO • SINGAPORE • SYDNEY • TOKYO
Newnes is an imprint of Elsevier

Newnes is an imprint of Elsevier
The Boulevard, Langford Lane, Kidlington, Oxford OX5 1GB, UK
225 Wyman Street, Waltham, MA 02451, USA

First edition 2014

British Library Cataloguing in Publication Data
A catalogue record for this book is available from the British Library

Library of Congress Cataloging-in-Publication Data
A catalog record for this book is available from the Library of Congress

ISBN: 978-0-12-801314-4

For information on all Newnes publications
visit our web site at books.elsevier.com

Printed and bound in United States of America
14 15 16 17 18   10 9 8 7 6 5 4 3 2 1

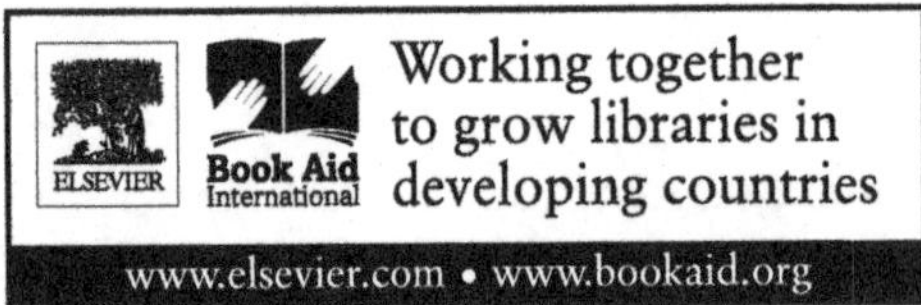

Contents

Introduction

Microcontrollers are computers on a chip. When they power up they start running a program from internal program memory, also called ROM for read only memory, or Flash. Microcomputers are found in appliances, toys, automobiles, and computer peripherals, such as a keyboard or mouse, and are finding their way in as support electronics for almost everything electronic from battery chargers to RADAR systems.

The Microchip PIC® microcontrollers have become the most popular choice for new designs based on their high speed, of up to 70 million instructions per second as of this writing; low cost, some under $1; and large number of interfaces like USB, Ethernet, and analog signals.

The C programming language, originally developed by AT&T Labs by authors Brian Kernighan and Dennis Ritchie, known as K&R C, became an international standard by ANSI in 1989, known as C89. A new standard derived from C defined the C++ language and was released in 1998. C++ has some complex language elements that make it impractical for use on a microcontroller as opposed to a desktop PC. C is the most commonly used language for programming microcontrollers.

C is in a category of computer languages called high order languages. High order languages use a tool called a compiler to convert the C text files to a machine readable file.

The first part of this book emphasizes the C language. Previous experience with a programming language will help but is not required. Formal definitions of the language elements are used and all areas of the language that apply to microcontrollers are covered in detail. Starting with Chapter 15, this book covers the PIC® microcontroller, its peripherals, and how to use those peripherals from C in detail. Prior knowledge of basic electronics to interface to hardware devices will help, but is not required to understand this book.

There are variations in the C language extensions between compiler vendors and between microcontroller families. Throughout the book are indications where there may be compatibility issues between compilers and/or processors. Each chapter will also supply hints for good programming practices, including documentation. Exercises and quizzes are provided for each chapter to help solidify the concepts learned. This book uses examples ready to compile using the CCS C compiler. It is the most popular compiler for the Microchip PIC® processors.

Trademarks:

PIC® MCU, dsPIC® DSC, and MPLAB® are registered trademarks of Microchip Technology, Inc., in the USA and other countries.

CHAPTER 1

C Overview and Program Structure

C Source Code

This is what C source code looks like:

```
/*   Chapter One sample C program for
     the CCS C compiler */

#include <e3.h>

void main(void) {
     int i;

     for( i=1; i<=10; i=i+1 ) {
             output_high(PIN_C6);          //Turn green LED on
             delay_ms(500);
             output_low(PIN_C6);           //Turn green LED off
             delay_ms(500);
     }
}
```

This program may look very cryptic to you now. When you have finished reading this book and doing the experiments, this program and much more complex ones will no longer seem cryptic. As you read the next few chapters, you can refer to this program to see how the topics relate to this program. For now, let's make some observations about the overall look of the program.

Comments

Comments help anyone (including you) who reads your code understand what it does. There are two styles.

Embedded C Programming. http://dx.doi.org/10.1016/B978-0-12-801314-4.00001-6

`/* */` Comments between `/*` and `*/` (can span multiple lines, but may not be nested)

`//` Comments between `//` and end of line (one line only).

The compiler ignores all comments. Both styles are shown in the sample program.

Program Structure

C programs are made up of compilation units, sometimes called translation units. A compilation unit is a set of files that are compiled together by a compiler. For most examples in this book we will be using a single compilation unit. A compilation unit is made up of global data and functions. A function is a callable group of code that in some other languages is referred to as a procedure or subroutine. Functions are made up of local data accessible only to the function and of statements within the function.

C Preprocessor Directives

An interesting feature of C is that it has what is referred to as a preprocessor. Think of the preprocessor as a tool that goes through the code first and makes some modifications to what is actually compiled. Preprocessor directives start with a # and occupy the entire line. They will be covered in more detail in Chapter 3. In the above example the **#include** directive causes whatever lines are in the file (e3.h) to appear at this spot in the code for the compilation.

For example, if you created a file named delay.inc and put in the file one line:

```
delay_ms(500);
```

then you could replace the two delay lines in the above program with **#include <delay.inc>** and the program would compile exactly the same. In the first step of compilation the compiler preprocessor would read the delay.inc file when it got to that line and replace the **#include** with **delay_ms(500);** where the **#include** were.

The preprocessor can be a powerful feature in C that can increase program readability, maximize code reuse, and significantly help in program maintenance.

As we examine the sample program shown, you see the first line is a preprocessor directive to include the e3.h file. It is very common for the first non-comment line in a program to include a file with various project- and/or hardware-specific definitions. The .h extension (for header) is frequently used for this kind of file. In our case all the declarations needed for the E3 hardware are in this include file.

Functions

Next we find a function definition for **"main"**. All programs must have exactly one function named **main()**. This is where the program begins execution. When referring to a function name in this book we will follow the name with `()` so it is clear that it is a function name. The `void` before the name indicates this function returns nothing and the `void` inside the `()` indicates this function gets nothing from the caller. The `{` and `}` are grouping symbols. All functions start and end with these symbols.

Functions will be dealt with in detail in Chapter 7; however, to lay a foundation for what they are, consider some examples of a function being used:

`x=sin(y);`	sin is a function with one argument and a return value
`x=sin(y*3.1415/180);`	the argument may be any expression
`x=180*sin(y)/3.1415;`	the return value may be used in an expression.

Declarations

The "`int i`" is a data declaration for the variable named with the identifier **i**. **int** indicates the variable is an integer. In this case `i` may only be used inside the `main()` function. If this line was above the start of the function (outside the function) then `i` could be accessed by other functions. The range that a variable is accessible from is referred to as the variable scope. Scope will be covered in more detail in Chapter 4.

Statements and Expressions

The `for` line is a statement. Statements are executed at run time. This particular statement has three expressions within. Expressions will be covered in Chapter 5 and statements in Chapter 6. The quick overview of the `for` statement is:

It executes the first expression once `i=1`
Repeats the following:
  Tests the second expression and exits the loop if false `i<=10`
  Executes the statement following the `)`
  The third expression is executed. `i=i+1`

In this example the four lines are executed 10 times with the variable `i` going from 1 to 10 and then, when 11, the loop stops because 11 <= 10 is false.

Expressions are some combination of constants, variables, operators, and function calls. Expressions always have a resulting value. Some simple examples of an operator are `+ - * /` and the very special `=`.

In our case since we have four statements to execute in the `for` we need to group them together with the `{` and `}`. This is called a compound statement. The braces may contain zero or more statements. Without those, only the `output_high()` function would be called in the loop 10 times. Then the other three lines would execute once afterwards.

Each of the four lines in our loop is a function call. These functions are not defined by the programmer but rather are functions built into the compiler. Function calls are recognized by the **(** following the function name. The expression(s) inside the **()** of a function call is the data passed into a function. These are called arguments in the call and parameters in the function.

In C a special case of a valid statement is any expression followed by a **;**. Note that just because it is valid does not mean it makes sense. For example, this is a valid C statement:

```
1+2;
```

However, it does not do anything. Some compilers might do the addition and that may take some time but nothing more is done. A good compiler will throw a warning on this line because the programmer might have made a typo.

A **;** with no expression before it is a special case of a statement called the null statement. It does nothing.

In C there is not an assignment statement as in some other languages, but rather an assignment operator, the **=**. Consider:

```
x=3;
```

This is an expression **x=3** consisting of a variable, operator, and constant. With the **;** it makes a statement. It always assigns the value on the right side (rvalue) to the variable on the left side (lvalue).

Time

The ms in **delay_ms** is milliseconds. Time units frequently used in programs are:

ns	nanosecond	0.000,000,001 seconds
us	microsecond	0.000,001 seconds
ms	millisecond	0.001 seconds

For example, there are 1 million microseconds in 1 second.

Typing Accuracy

Typing accuracy is very important when creating C source code. A punctuation mark, either typed by mistake or omitted, can cause a lot of head scratching because your program will not compile. The compiler sees **exactly** what you type.

For example, if the **{** was missing on the for line then the compiler would trigger an error when it got to the **}** line four lines down from where the actual error was.

The **;** that follows many statements and declarations is important to help the compiler to know when a definition or statement ends. It is never used at the end of a preprocessor directive that starts with **#**. Missing or extra **;** or **{** can create confusing error messages. A good C editor will highlight matching **{ }** and **()** and as well as highlight syntactical elements to prevent errors as you type.

Text Formatting

Formatting white spaces such as spaces, tabs, carriage returns, etc., are ignored by the compiler. Formatting makes code readable to us. White space, resulting from laying out a program so it appears better organized, is a good thing. This comes from using spaces, tabs, and blank lines. Use tabs for indentation instead of several spaces. The number of spaces per tab is usually adjustable. Three spaces per tab work well. Notice the lines inside the above function are indented and the lines inside the loop are further indented. Indentation and other white space are optional, but highly recommended. There is no right or wrong as far as the compiler is concerned. Some companies will have companywide coding standards that will specify indentation, comments, maximum function size, and other readability items.

Compatibility Notes

The **//** comment is a C++ construct not supported by all C compilers.

Most C Compilers are sensitive to case. For example, **Output_High()** would not be recognized but **output_high()** is. By default, the CCS C compiler is not case sensitive. To make it case sensitive, you must use a **#case** preprocessor directive.

Built-in functions like **output_high()** and **delay_ms()** are not in the C standard. These are unique to the CCS C compiler.

Summary

- Programs are made up of one or more compilation (or translation) units.
- Compilation units have preprocessor directives that are resolved before anything else.
- Comments and most white space are ignored.
- A compilation unit is a file with some number functions and global data declarations in any order.
- Functions have local data declarations and statements all enclosed in **{ }**.
- Functions may return data and the caller may pass data known as arguments to the caller and parameters to the function.
- Groups of statements may be enclosed in **{ }** to make a compound statement.
- Some statements have expressions within them.
- A statement may be any expression followed by a **;**.
- Expressions are made up of constants, variables, operators, and function calls and always evaluate to some value.

Exercise 1-1

Objective: Gain a basic understanding of how to use the compiler and prototyping board.
Requires: E3 module, USB cable, PC.

Steps/Technical Procedure	Notes
1. For help installing and running the compiler, as well as setting up the hardware, consult Appendix B.	
2. Create a source file. **File > New > Source** Enter Ex1-1; the IDE will add .c file name extension.	
3. Type in the program shown earlier in this chapter and save it via **File > Save**	
4. Compile your code. Click "Compile" on the menu bar at the top of the screen. The Compile menu ribbon will appear. Click on the Compile icon on the Compile menu ribbon. Your source code will (hopefully) be compiled successfully. The Output window will appear, indicating how the compile process turned out. If you got 0 errors, that is what we wanted, so celebrate!	
5. Connect the E3 board to a USB cable and plug into the PC. If Windows detects the E3 as a new device and needs to find drivers, point it to the "c:\program files\PICC\USB Drivers" directory. "program files" may have a different name on your computer. Note that the E3 board is powered by the USB cable. The E3 board has a program preloaded onto it to allow for downloading user programs over the USB.	
6. From the IDE Compiler ribbon select PROGRAM CHIP and then E3 BOOTLOAD.	
7. At this point, if all went well, you should have a green LED blinking once per second on the E3 board. A program window (Serial Port Monitor) will popup. For now, close this window. It will be used in later exercises.	
8. Modify the program so the LED blinks once every 5 seconds. Compile Download Verify it works.	

Exercise 1-2

Objective: Better understand the use of the tools.
Requires: E3 module, USB cable, PC.

Steps/Technical Procedure	Notes
1. To show error handling, change both of the PIN_C6 to PIN_C66 and compile. Notice the error messages indicate PIN_C66 is not known. Double-click on one of the error messages to move the cursor to the spot where the error was detected. Correct the program and recompile.	
2. Remove the "void" between the **(** and **)** in the main() function definition, and compile. This time notice there were no errors, but one warning. The compiler uses warnings in cases where the compiler was able to do what you asked for but suspects you may have done something wrong. In this case the compiler is concerned you forgot to list the function parameters. Inserting void is a positive indication there are no parameters. Correct the program and recompile.	
3. Use **VIEW > DATA SHEETS > Other PDF's > E3 Schematic** to view the board schematic and determine why we specified PIN_C6 to light the green LED.	
4. Modify the program to blink the red LED. Verify it works.	
5. Figure out a way to modify the program to find out if the LED is on when the pin is made high or low.	
6. On the compile ribbon, click on the C/ASM button to view the mixed C and assembly language. This is called the list file, with a .LST extension. Count the assembly instructions in the program. Calculate the assembly instructions per C line (not including preprocessor directives, comment and blank lines). This statistic is sometimes used to determine how efficient a compiler is. Instead of using C lines, recalculate using C statements. Some believe statements are a better indicator of the C to ASM ratio.	
7. Modify the program so the LED blinks once every 5 seconds. Compile Download Verify it works.	

Quiz

(1) As an attempt to reduce white space, which of the following programs is valid?

(a)
```
#include<e3.h> void main(void){inti;for(i=1;i<=10;i=i+1){
output_high(PIN_C6);
delay_ms(500);output_low (PIN_C6);delay_ms(500);}}
```

(b)
```
#include<e3.h> void main(void){int i;for(i=1;i<=10;i=i+1){
output_high(PIN_C6);
delay_ms(500); output_low(PIN_C6);delay_ms(500);}}
```

(c)
```
#include<e3.h> void main(void){int i;for(i=1;i<=10;i=i+1){
output_high(PIN_C6);
delay_ms(500);output_low (PIN_C6);delay_ms(500);}}
```

(d) All are valid

(e) None are valid

(2) Which of the following statements is valid?

(a) `5;`

(b) `1+2;`

(c) `{1;2;3;4}`

(d) All are valid C

(e) None are valid C

(3) How many times will this statement loop?

```
for(i=1; i>=10; i=i+1)
```

(a) 11

(b) 9

(c) 1

(d) 0

(e) Infinite

(4) How many times will this statement loop?

```
for(i=1; i<=10; i=i+3)
```

(a) 0

(b) 2

(c) 3

(d) 4

(e) Infinite

(5) What C elements can be found in this line of code?

```
delay_ms(50*n);
```

(a) Function declaration
(b) Expression
(c) Statement
(d) Argument
(e) All of the above
(f) b, c, d
(g) c, d

(6) Of the following comments, which ones are valid C?
(a) int x // holds current x position;
(b) process(data); /* removes symbols +-*/ in the data */
(c) // temporary>> // int dummy;
(d) All are valid
(e) None are valid

(7) What should the compiler do with this line?

```
for(1;2;3)
```

(a) Generate an error
(b) Generate a warning
(c) Both a and b
(d) Neither a nor b

(8) A single expression may also be:
(a) A statement
(b) A declaration
(c) A parameter
(d) Functions
(e) A translation unit

(9) A file q1.inc has the following lines:

```
output_high(PIN_C6);
delay_ms(500);
output_low(PIN_C6);
delay_ms(500);
```

With the following code, how many times will the LED blink?

```
for(i=1; i<=3; i=i+1)
#include <q1.inc>
```

(a) 1
(b) 2
(c) 3
(d) 4
(e) None

(10) A file qq1.inc is created with just four characters being **void**. When an attempt is made to compile the following, how many errors are generated?

```
#include <qq1.inc>
main(
#include <qq1.inc>
) { }
```

(a) None
(b) Three, two void with nothing after and one **(** with no **)**
(c) Two, only the void lines
(d) One, only the main line
(e) Four, the three from b and missing statements between **{** and **}**

CHAPTER 2

Constants

Constants have already been briefly introduced. C allows for constants to be expressed in a number of different ways. In some cases there are different ways to express the exact same number, and in other cases the way a constant is expressed indicates the constant's type. Types will be covered in more detail in the next chapter. Types dictate the organization of an item in memory. For example, how many bytes the item takes.

Bits, Bytes, Etc.

Bits

One bit in memory or a register can represent one of two possible states, "0" or "1."

In the world of digital electronics, it is convenient to build circuit elements that have two states: off/on, active/inactive, or low/high. These states can be represented by "0" or "1" (see Figure 2.1).

The exact voltage ranges that represent 0 and 1 vary depending on the logic supply voltage and the integrated circuit logic chip family used (TTL, CMOS, etc.). The choice of using binary 0 to represent 0 V is arbitrary. Positive logic is shown above. It can be done the opposite way, which is called negative logic.

Nibbles

A nibble consists of 4 bits and can represent 16 possible states. A nibble is typically the upper or lower half of a byte (most significant or least significant nibble).

Bytes

A byte consists of 8 bits and is said to be 8 bits wide. An 8-bit microcontroller moves bytes around on an 8-bit data bus (8 conductors wide).

Embedded C Programming. http://dx.doi.org/10.1016/B978-0-12-801314-4.00002-8

POSITIVE LOGIC	
Binary Number	Voltage
0	0 volts (approximately)
1	5 volts (approximately) in a 5 volt system

Figure 2.1: Positive logic.

Memory Sizes

Memory size is most frequently expressed in terms of the number of bytes. Sometimes it will be expressed in terms of the number of words, where the word size is specific to the way memory is organized in the machine. For a PIC® MCU, frequently RAM is expressed in bytes and program memory as words because the program memory may be 12, 14, 16, or 24 bits wide.

Because addressing is done in binary, large memory sizes are also expressed as a power of 2. For example, instead of 1000 bytes it will be expressed in terms of 1024 bytes (2^{10}). Here are the common abbreviations that apply ONLY to memory:

1K = 1 kilobyte = 1024 bytes
1M = 1 megabyte = 1024 × 1024 or 1,048,576 bytes
1G = 1 gigabyte = 1024 × 1024 × 1024 or 1,073,741,824 bytes.

Syntax of C Constants

Binary

A binary number with more than one bit can represent numbers larger than 1. How much larger depends on the number of bits or digits. An 8-bit binary number (byte) can represent 256 possible numbers (0–255).

A 16-bit binary number can represent numbers from 0 to 65,535. If we use a byte to transmit information, we can transmit 256 possible combinations, enough to represent the 10 decimal digits, upper- and lowercase letters, and more. A commonly used code used to represent these characters is called ASCII (American Standard for Information Interchange).

Binary numbers are based on powers of 2. The value of bit 0 is $2^0 = 1$ if it contains a 1, or 0 if it contains 0. The value of bit 1 is $2^1 = 2$ if it contains a 1, or 0 if it contains 0. The value of bit 3 is $2^3 = 8$ if it contains a 1, or 0 if it contains 0, and so on.

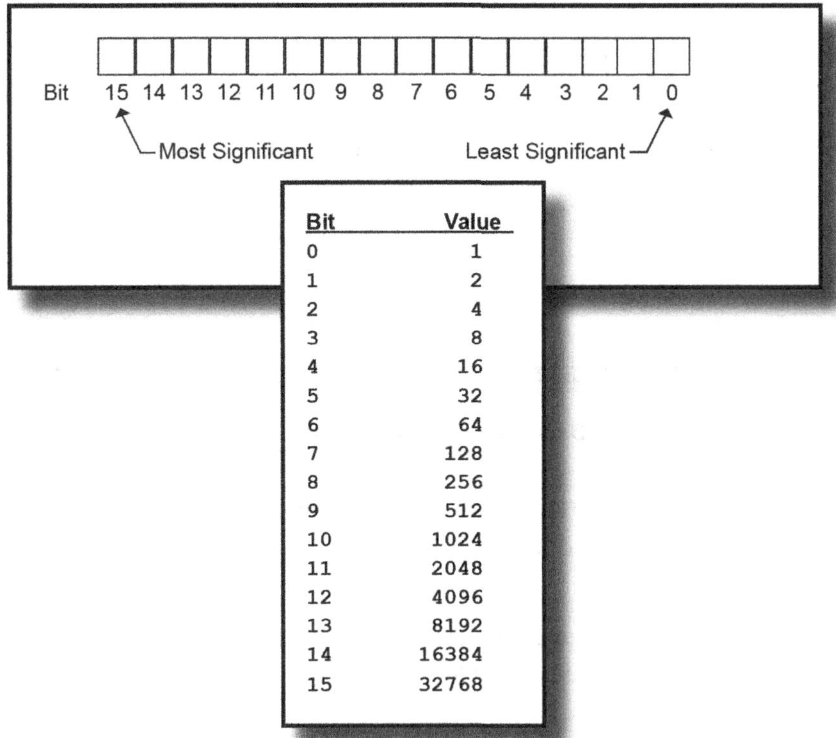

Bit	Value
0	1
1	2
2	4
3	8
4	16
5	32
6	64
7	128
8	256
9	512
10	1024
11	2048
12	4096
13	8192
14	16384
15	32768

Figure 2.2: Binary bit value in decimals.

For a 16-bit binary number, bit 0 is the least significant bit, and bit 15 is the most significant bit. Figure 2.2 shows the value of each bit position if it contains a "1": All PIC- related documentation numbers the least significant bit as bit 0. This is a frequent but not universal convention for microprocessors.

The value of a binary number contained in a 16-bit variable would be determined by multiplying the contents of each bit by the value of each bit (see Figure 2.3).

Counting up in binary goes like this:

0000
0001
0010
0011
0100
0101
etc.

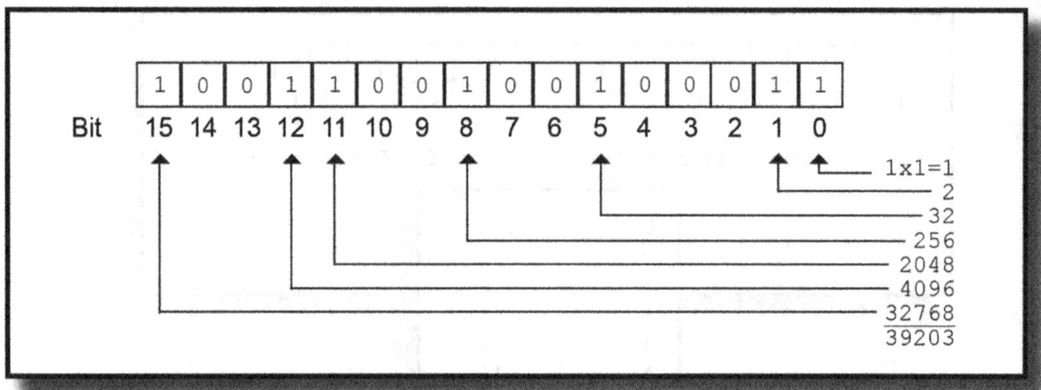

Figure 2.3: Binary-to-decimal conversion example.

Think of this number system in terms of a car odometer where each digit only has a 0 and 1. When the 1 flips over to 0 the next digit up flips. The way a number is represented is called the radix.

In C a binary constant starts with a **0b** followed by each binary digit. For example, to assign a decimal 6, or binary 00000110, to the variable x do this:

```
x=0b00000110;
```

Decimal

Decimal numbers are the way most people think of numbers. There are 10 digits in each decimal position 0–9. In C we express decimal simply like this:

```
x=123;
```

Types will be dealt with later in this book; however, we will note here that the compiler will treat the above 123 as a 1-byte constant since 123 fits into a byte. If for some reason the programmer wants it to be treated as a 2-byte number then an **L** (for long) is appended to the number, like this:

```
x=123L;
```

The **L** is not required for numbers that are already too large to fit in a byte (like 1000).

A number can also have a **U** after it to indicate it is unsigned.

Signed Integers

The binary numbers shown above are unsigned numbers. An 8-bit memory cell can have a number from 0 to 255. It is sometimes helpful to represent negative numbers in C. When doing so, one binary position is used for the sign. 0 is positive and 1 is negative. The rest of the binary positions are in what is referred to as a ***2's complement format***. The 2's complement format has an arithmetic advantage over what might seem to be a simpler scheme. A −2 in 2's complement form at looks like this:

1 1 1 1 1 1 1 0

When adding 1 to this, the result is:

1 1 1 1 1 1 1 1

And this is the 2's complement representation for a −1. That is the preferred way it should work. Add 1 to −2 and get −1. If adding another 1 (think of the odometer) the result is:

0 0 0 0 0 0 0 0

Just as any mathematician would expect.

Think of the odometer starting out at all 0s. If wanting to represent a −1, then move it backwards 1 to result in all 1s.

In C, signed integers are represented like this:

```
x=-123;
```

Hexadecimal

The hexadecimal number system (or radix) has 16 digits, 0–9 and then A–F. Again, think of this in terms of an odometer with 16 digits in each spot. The reason for hexadecimal is that it is very useful for programmers to understand numbers in terms of their binary representation; however, long binary strings of digits are hard to remember and communicate. Unlike the more commonly used (in the real word) decimal radix, hexadecimal has a direct relationship to binary. Every hexadecimal digit is exactly 4 bits in binary. A four-digit hexadecimal number represents cleanly 16 values (see Figure 2.4). Many programmers will have memorized the 16 binary patterns that correspond to the 16 hexadecimal digits. They can therefore visualize the binary pattern in a given hexadecimal number. For example, for someone expecting data over a communications channel of AA, and they are getting 55, it is immediately clear one bit was lost and then the data is the same.

Hex is sometimes used in this book to represent addresses in memory and is sometimes used to represent bytes of data. Using hex is not difficult. All you need is a little practice.

One byte requires two hex digits. Note that the digits representing a byte are sometimes shown in groups of two or four and that the most significant hex digit is on the left.

Hex numbers are denoted by 0x in this book. Some of the more recent microchip literature uses "h" to designate a number as hexadecimal (this is not valid in C).

A summary of the relationship between binary, hex, and decimal for some selected numbers is shown in Figure 2.5.

Hexadecimal	Binary	Decimal
0	0000	0
1	0001	1
2	0010	2
3	0011	3
4	0100	4
5	0101	5
6	0110	6
7	0111	7
8	1000	8
9	1001	9
A	1010	10
B	1011	11
C	1100	12
D	1101	13
E	1110	14
F	1111	15

Figure 2.4: Hexidecimal conversion chart.

	Binary	Hex	Decimal
Nibbles	0000	0	0
	0001	1	1
	0101	5	5
	1111	F	15
Bytes	0000 0000	00	0
	0000 0001	01	1
	0001 0000	10	16
	1111 1111	FF	255
Two Bytes	0000 0000 0000 0000	0000	0
	0000 0000 0000 0001	0001	1
	0000 0001 0000 0000	0100	256
	1111 1111 1111 1111	FFFF	65535

Figure 2.5: Radix conversion example.

Hexadecimal 0xFFFF (the very top of the program memory address space in some microcontrollers) is much easier to write or remember than either 1111 1111 1111 1111 or 65,535.

In C a hex constant starts with a **0x** followed by each hex digit. For example, to assign a decimal 26, or hex 1A, to the variable x do this:

```
x=0x1A;
```

Octal

The octal radix has eight values for each digit (0–7). Like hexadecimal this number system shares a direct relationship to binary. There are 3 bits in each octal digit. Early computer systems used octal before hex became more popular because you can get more bang for each digit (see Figure 2.6). Although the octal radix is hardly ever used anymore, it is mentioned because C gives it a prominent place in the language syntax.

In C an octal constant starts with a 0 followed by each octal digit. For example, to assign a decimal 10, or octal 12, to the variable x do this:

```
x=012;
```

Note that it is a zero, not O as in orange after the **=**. Be very careful; even though you do not intend to use octal, if you put a leading 0 on a number you have just converted your intended number to octal. In the above line of code, **x** would represent the number of fingers you have, not the number of eggs in a carton.

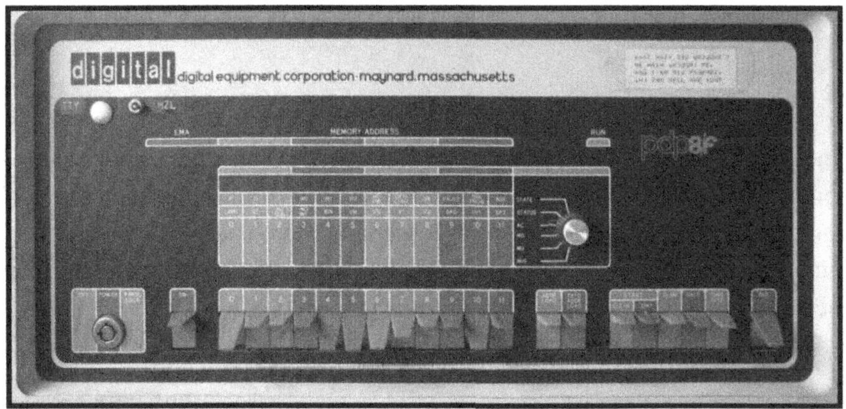

Figure 2.6: Front panel of a DEC PDP-8.
The levers of the DEC PDP are used to set up a binary machine code instruction, then the load button loads the instruction into the next RAM location. Notice how the colors are grouped into 3 bits for easy octal entry. These computers were popular when C was created. *(Photograph: PDP-8/F provided by Herb Johnson, http://www.retrotechnology.com/.)*

Floating Point

The numbers mentioned above are all whole numbers, no decimal point. A floating-point number is a number with a decimal point that moves over a great range. A simple floating-point number is 12.34. Internally the number is saved as 1234×10^{-2}. The two parts are referred to as the mantissa and exponent. A given compiler will have a maximum range for each. On the smaller PIC® devices, the CCS C compiler has 24 bits for the mantissa and 6 bits for the exponent. With 2 bits for the sign of each, this totals 32 bits. This means the following numbers can be represented well:

1.23	123×10^{-2}
0.000000000000123	123×10^{-15}
12,300,000,000,000,000	123×10^{14}.

But a number like this:

1,200,000,000,000,003	$1{,}200{,}000{,}000{,}000{,}003 \times 10^{0}$

will turn into:

1,200,000,000,000,000	120×10^{14}

because the mantissa is not large enough for that many significant digits.

Because there is not always a clean translation to decimal when comparing floating-point numbers, use less than and greater than rather than equals and not-equals whenever possible.

In C there are three ways to express a floating-point number, as follows:

```
x=1.23;
x=1.23E5;     // same as 123,000.0
x=5F;         // same as 5.0.
```

Fixed Point

A fixed-point number has a decimal point, but the decimal point is fixed. For example, if dealing with money, there is no need for more than two decimal points. The number can be saved like an integer and tell the compiler to always remember there is a decimal point before the last two digits. There is no special designation for constants that are fixed-point, the variable will have the fixed-point designation. The only form allowed is:

```
x=1.23;
```

Characters

Characters include the digits (0–9) and letters (A–Z) as well as punctuation and special symbols that are used in programming. Each character that can be represented in C corresponds to a number from 0 to 255. ASCII is a standard for characters that may be represented by 8 bits.

Table 2.1 C character escape sequences.

\n	Line feed—same as \x0a
\r	Return—same as \x0d
\t	TAB—same as \x09
\b	Backspace—same as \x08
\f	Form feed—same as \x0c
\a	Bell—same as \x07
\v	Vertical space—same as \x0b
\?	Question mark—same as \x3f
\'	Single quote—same as \x22
\"	Double quote—same as \x22
\\	A single backslash—same as \x5c

There are 256 ASCII characters. The ASCII chart is given in Appendix A. For example, the letter A is stored in memory as 65. A character constant is enclosed between single quotes. C is somewhat loose in typing (translating items of various types, nothing to do with what your fingers do on the keyboard) so you can always use 65 or 'A' interchangeably in your code. An assignment looks like this:

```
x='A';
```

There is a special way to specify characters that may not have a keyboard representation, called an escape sequence. The `\` is used to indicate this special format. If you want a `\` character you must use two in a row like this: `\\`. In order to just specify the hex value for the character you do this: `'\x41'` (this is the same as `'A'`). For an octal representation use: `'\101'`. There are a number of additional special characters in Table 2.1.

Note that the backslash notation must be used to get a backslash character itself as well as the single and double quotes in some cases. For example, the following lines are the same and in each case a single byte value is written to c:

```
c='\r';
c='\0x0D';
c=13;
```

String of Characters

C provides for a string of characters to be represented in memory. In this case there can be any number of characters and the end is determined by a null character `'\000'` at the end of the string. For example:

`"ABCD"` memory contents | 41 | 42 | 43 | 44 | 00 |

Using the double quotes, we have specified a string of characters. In this case there are four characters in the string and five characters in memory (the last one being 0). This string requires 5 bytes of memory.

Strings can have within them special characters, for example:

```
"A\0x42\111D"
```

memory contents	41	42	43	44	00

In this case the string is the same 5 bytes. The B and C were represented by hex and octal. It is very common to use the **\r** (return) and **\n** (line feed) in C strings. The Windows file system uses these two characters to identify the end of a line in a file. A string that represents a single line in a file and on the screen of most terminal programs might look like this:

```
"Line One\r\n"
```

memory contents	4C	69	6E	65	20	4F	6E	65	0D	0A	00

Another interesting C feature is if two strings appear in code next to each other with optional white space between them, then the two strings are treated as one big string. For example:

```
"ABCD" "EFGH"
```

will give you the same thing as:

```
"ABCDEFGH"
```

memory contents	41	42	43	44	45	46	47	48	00

Notice in this case the 0 after the D is discarded so for both of the above the size in memory is 9 bytes.

This feature can be used when dealing with very long strings that you may want to break up onto multiple lines. It also helps in macros, which we will cover in a later chapter.

True and False

A C expression evaluates to a numeric value. Expressions using relational operators evaluate to a value of either **TRUE** (1) or **FALSE** (0). Relational expressions are often used within **if** and **while** statements.

```
if (a<b) //compare a to b
```

If the expression evaluates as **TRUE**, the statement(s) following this line of code are executed. If the expression evaluates as **FALSE**, the statement(s) following this line of code are not executed. The relational expression **a<b** will evaluate to a 0 or 1. The **if** statement simply treats any nonzero value as a true and a zero value as false. The CCS C compiler predefines in the device header file two identifiers TRUE and FALSE that sometimes will be useful when dealing with relational expressions. The C standard does not define TRUE and FALSE.

Const

We have already used a simple data declaration like this:

```
int i;
```

A symbolic constant may be defined using the key word **const**, which is a qualifier that can be applied to a declaration.

```
const  int  LEVEL = 10;   // defines integer type constant named
                          // level with a value of 10
```

After the symbolic constant is defined, it is referred to by name in the program. In this case, **LEVEL** can never be changed in the code, it is always 10. The more common way to define constants like this is:

```
#define LEVEL 10
```

This preprocessor directive will replace all occurrences of **LEVEL** with 10. This will be covered in more detail in the next chapter.

One advantage of the **const** method is the identifier type (in this case **int**) is positively associated with the identifier along with the value. Frequently programmers will make **const** identifiers all uppercase.

Tri-Graph Sequences

The C language has some dated elements; however, because they are part of the specification we must work around them (like octal). Not all computer keyboards have all the special characters used in the C language. For example, if your keyboard looks like the one in Figure 2.7, the tilda ~ is not on the keyboard. The solution to this is what is called a tri-graph sequence. It is a set of three characters that are always processed before anything else (like a preprocessor). A full list of the tri-graph sequences is in Table 2.2. The sequence always starts with two question marks and herein lies the problem. If you have two question marks in a row in your code (even in a quoted string) you may need to work around tri-graphs. To get two question marks you need to put six in the code. The following is an example of two lines of code, one with and one without tri-graphs.

```
x=~ y;
x=??-y;
```

Figure 2.7: ASR-33 terminal.

The compiler accepts three-character sequences instead of some special characters not available on all keyboards, as follows:

Table 2.2

<table>
<tr><th>Sequence</th><th>Same as</th></tr>
<tr><td>??=</td><td>#</td></tr>
<tr><td>??(</td><td>[</td></tr>
<tr><td>??\</td><td>\</td></tr>
<tr><td>??)</td><td>]</td></tr>
<tr><td>??'</td><td>^</td></tr>
<tr><td>??<</td><td>{</td></tr>
<tr><td>??!</td><td>|</td></tr>
<tr><td>??></td><td>}</td></tr>
<tr><td>??-</td><td>~</td></tr>
</table>

Compatibility Notes

Fixed-point data is not supported in many C compilers. Those that support it will usually have a very unique way of specifying the data. This will usually be tied to the way it is implemented in hardware. Some DSP processors have built in fixed-point math units.

The **const** qualifier, according to the C standard, simply makes an object read-only. Different compilers will implement this in different ways. In the CCS C compiler, **const** is used to force data into the program memory. The method by which the data is saved in memory is such that it is easiest to access. In cases where the programmer is depending on the data being in a certain format in memory, the **const** may not work. Another qualifier, **rom**, may be used in the CCS C compiler, that will still use program memory but in a format that can be used with all the C operators, including pointers.

Design Documentation

When program constants that have identifiers (like **MAX_NUMBER_OF_ENTRIES**) are used in a program and formal documentation is required, a document is usually generated that will list all these constants and describe what they are for. Comments in the code to describe these constants will help to better understand the code and aid in the generation of this documentation.

Tools are available that can extract the comments associated with constants from the code. Here is an example:

```
#define  PIXELS_PER_INCH  120     // Number of pixels in one inch on the
                                  // screen
#define  MAX_ENTRIES              // The maximum number of device entries
```

Summary

- Bytes have 8 bits, and nibbles have 4 bits.
- The binary radix (0b prefix) has 2 values for each digit and has 2-digit numbers.
- The octal radix (0 prefix) has 8 values for each digit and has 8-digit numbers.
- The decimal radix (no prefix) has 10 values for each digit and has 10-digit numbers.
- The hexadecimal radix (0x prefix) has 16 values for each digit and has 16-digit numbers.
- Signed numbers are in a 2's complement format making common math easy.
- Floating-point numbers have a very large range and a limited accuracy.
- Floating-point data has a mantissa and exponent and those sizes determine the number range and accuracy.
- Fixed-point numbers have a precise accuracy to a specific number of digits and a limited range.
- Characters are specified with single quotes and are encoded to numbers using the ASCII translation charts.

- Special character sequences may be used to specify a specific encoding or to use common non-printable characters using the \ lead character.
- Strings of characters are represented in C with double quotes and always have a 0 terminator in memory.

Exercise 2-1

Objective: Gain an understanding of C constants by use of the USB interface on the E3 module to send data to the PC screen.

Requires: E3 module, USB cable, PC.

Steps/Technical Procedure	Notes
1. Recall the Serial Port Monitor program closed every time a program is loaded. Now let us start to use it. The following function call will output formatted data to that screen from the running program: `printf("x is %u \r\n", x);` The **printf** will be covered in detail later. For now, understand the string in double quotes is sent to the console and whenever a **%** is seen, a variable found as the next argument is formatted and sent to the console. The **u** after the **%** means unsigned format. You can use **%d** for a signed format, **%x** for hexadecimal, and **%c** for a character.	
2. Write a program using the for statement previously used to output the numbers 1–20 on the console.	
3. Update the program to display the numbers in hex.	
4. Now change the program to loop from 65 to 90 and display the number as a character.	
5. The **printf** allows for multiple format specifies **(%)** in the same call. Each **%** expects another argument in the function call (separated by commas). Change the program to loop from 33 to 52 and display the numbers as unsigned, hex, and character.	
6. Remove the loop and with a single **printf** send the string "Hello World" to the console; however, for each character in the string specify the character using its hex representation.	
7. Use **printf** to show the hexadecimal representation in memory for a −100.	

Quiz

(1) How many bytes of memory are required to hold the following binary number? 1001011001
- (a) 10
- (b) 1
- (c) 1.2
- (d) 2
- (e) None, binary numbers cannot be put into bytes

(2) One hex digit can be described in which of the following terms?
- (a) Two decimal digits
- (b) Byte
- (c) Nibble
- (d) One octal digit
- (e) Three binary digits

(3) The binary value 0011 in hex is 3. What is 00110011 in hex?
- (a) 303
- (b) 33
- (c) 6
- (d) 1111
- (e) 3F

(4) The hex number 1F plus one is what?
- (a) 1G
- (b) 1F1
- (c) 1E
- (d) 21
- (e) 20

(5) What does the following line show on the console?

```
printf("Value is %u", 081);
```

- (a) Value is 81
- (b) Value is 081
- (c) Value is 65
- (d) Value is 129
- (e) Nothing, the line will not compile

(6) Given the following lines, which line will not print the same as the others?
- (a) `printf("Value is %u \r\n", 10);`
- (b) `printf("Value is %u \r\n", 0b1010);`

(c) `printf("Value is %u \r\n", 012);`
(d) `printf("Value is %u \r\n", 0x0A);`
(e) None, all will print the same

(7) How many bytes of memory are required to store the following constant string?
"three"
(a) 0
(b) 3
(c) 5
(d) 6
(e) 7

(8) How many bytes of memory are required to store the following constant string?
"\101\0x432\f"
(a) 0
(b) 3
(c) 4
(d) 5
(e) 6

(9) How many of the 256 ASCII characters cannot be represented in a C string?
(a) 0
(b) 1
(c) 31
(d) 128
(e) 159

(10) What error is in the following C code?
(i)
```
const int a=10;
int b;
int c;
c=b>a;
```
(a) A is not capitalized
(b) There is a semicolon on a const line
(c) A relational operator (>) can not be used with an assignment operator (=)
(d) All of the above are errors
(e) There are no errors

CHAPTER 3

Preprocessor Directives

C programs are processed by the compiler in two distinct steps. The first pass is a pre-processor step. The preprocessor directives, which begin with **#**, may affect compiler settings or may cause textual replacements. Be aware that the preprocessor variables (identifiers) are not the same as normal C variables. When the preprocessor is done, there will be no preprocessor directives or identifiers left for the normal processor.

A summary of the popular preprocessor directives follows:

Standard Preprocessor Directives

#define id text

id is the name you wish to define
text is the replacement text

In use, text replaces id everywhere it appears as the program is compiled. The **#define** performs a simple text replacement. Figure 3.1 shows examples of **#define**s and their use. These **#define**s are called macros.

The first define is a classic way to define a constant that is either important and subject to change, or appears in many places in the code. Notice the comment after the number 15. The comments are removed before the preprocessor gets the code. If that was not the case, the `; i=i+1 ) {` would be part of the comment and ignored.

The second define actually uses a define from e3.h (PIN_C6) that was **#define**d as 31766. All this gets resolved at preprocessor time.

The third and fourth defines show an example of a statement in a define. This also shows how a define can use another define.

The sixth define shows a formula in a define. It is always good practice to enclose formulas in `( )`. Failure to do so can cause unexpected results. For example, consider a define and its use:

```
#define MAX  100
#define HIGHEST  MAX-1


HIGHEST*3
```

Embedded C Programming. http://dx.doi.org/10.1016/B978-0-12-801314-4.00003-X

```
#include <E3.h>

#define NUMBER_OF_BLINKS   15        // Total number of blinks
#define GREEN_LED          PIN_C6
#define GREEN_LED_ON       output_low(GREEN_LED)
#define GREEN_LED_OFF      output_high(GREEN_LED)
#define BLINKS_PER_SECOND  1
#define BLINK_DELAY        (BLINKS_PER_SECOND*1000/2)

void main(void) {
    int i;
    for( i=1; i<=NUMBER_OF_BLINKS; i=i+1 ) {
        GREEN_LED_ON;
        delay_ms(BLINK_DELAY);
        GREEN_LED_OFF;
        delay_ms(BLINK_DELAY);
    }
}
```



```
void main(void) {
    int i;
    for( i=1; i<=15; i=i+1 ) {
        output_low(31766);
        delay_ms((1*1000/2));
        output_high(31766);
        delay_ms((1*1000/2));
    }
}
```


Figure 3.1: Example of #defines and usage.

In this case after the preprocessor you get 100−1*3, and that is not the same as (100−1)*3.

The preprocessor does not know C, it is doing only a text replacement. This can make analyzing error messages challenging. Consider a typo made in the sixth define where a **;** was typed instead of a *****. In this case the error would be flagged on the line:

```
for( i=1; i<=NUMBER_OF_BLINKS; i=i+1 ) {
```

That line appears correct. However, this line uses the sixth define where the typo occurred in the defined text replacement.

Notice we used a lowercase `i` for the variable and all uppercase for the **`#define`**s. This is sometimes done as a style or coding standard in order to recognize **`#define`** symbols.

The preprocessor directive takes up the entire line starting with the **#**. It is sometimes needed to use multiple lines for a preprocessor directive. To continue a directive on the next line, the \ symbol is used. For example:

```
#define BLINK_LED output_high(PIN_C6); \
                  delay_ms(500);  \
                  output_low(PIN_C6);
```

There is a special form of a **`#define`** that makes it look like a function call. These are called function-like macros or macros with arguments and are discussed in detail in Chapter 13.

There are some macros that are predefined by the compiler. For example:

`__date__` and `__time__`

These have the compile date and time predefined.

#include <filename> or #include "filename"

#include was covered in Chapter 1. In summary, the contents of the file are used at this point in the code. The first form with **< >** will first search in the predefined include file directories for the file. The second form with **" "** will first search for the file in the project directory. Usually the **< >** is used for compiler-supplied include files and the **" "** is used for your project-related files. In either form it is allowed to fully specify the file like this:

```
#include "c:\users\john\projects\includes\myboard.h"
```

This does make it harder to move project files to another directory, however.

The list of directories used to search for a file can be specified in the IDE, on the command line, or in a *.ini* file associated with the compiler.

A device file (for example 16F887.h) is almost always **#include**d in a project. For the examples in this book there is a device include at the top of e3.h. This shows it is allowable for a include file to #include other files.

#ifdef #ifndef #else #endif #undef

#ifdef is one form of conditional compilation. The lines between **#ifdef** and **#endif** will be ignored by the compiler unless the identifier has previously been defined with **#define**. Consider the following:

```
#ifdef  DEBUG
printf("Value=%u",value);
#endif
```

Many of these could be sprinkled throughout a program and the extra prints would only happen if a single line like this were added:

```
#define DEBUG
```

Notice there is no text after the identifier. In this case none is needed. **#else** is used like this:

```
#ifdef  OLD_TRANSDUCER
reading=rawdata/860;
#else
reading=rawdata/942;
#endif
```

It is important to understand the **#ifdef** is evaluated at compile time, not run time. Only one of the **reading=** lines will be compiled and put into the chip memory. The decision is made when the code is compiled. This is a powerful tool that allows one source code base to be easily reconfigured for multiple applications. The **#ifndef** is true when the identifier is not defined. The **#undef** is used to un-define an identifier that was previously **#define**d.

#if #else #elif #endif

The **#if** works like the **#ifdef**; however, instead of checking to see if an identifier is defined, it checks to see if the expression is **TRUE** (or nonzero). The identifiers in the expression must be macros not C variables. Here is an example that also uses the optional **#elif** (else if):

```
#if   REVISON==1
output_low(PIN_B0);
#elif  REVISION==2
output_low(PIN_B1);
#elif  REVISON==3
output_low(PIN_C0);
#endif
```

Expressions will be covered in more detail in Chapter 5; for now be aware the **==** operator is a test of equality. The **=** is exclusively used as the assignment operator.

#error

This directive is used to force an error. The text is output in the error message. This can be useful to find out if an area of code is being compiled. With **#if** and **#include**, sometimes code you think should be compiled is not being compiled. Put a **#error** in the code that should be compiled and if no error is thrown then the code is being ignored. One cause of this is a missing **#endif**. A missing **#endif** in a **#include** file can be very difficult to find since the code being ignored is after the **#include**, not even in the same file that has the error.

The most common use is to do something like the following:

```
#if  BUFFER_SIZE>256
#error  Buffer size is too large for the int8 index
#endif
```

This stops compilation if some combination of macros is specified in an illegal way.

For some compilers, like the CCS C compiler, the text after the **#error** is evaluated by the preprocessor.

For example, the following line in the #define example above will throw an error that looks like the line below if:

```
#error  Green=GREEN_LED
*** Error 119 "test.c" Line 16(7,20):  Green=31766
```

#nolist #list

The **#nolist** is used to tell the compiler to stop putting lines into the list output file (.lst). The **#list** resumes the normal operation. The LST file is created by the compiler and will show the assembly code generated for each C source line. This can be used to prevent lengthy comments or data definitions from taking up space in the list file. The compiler device header files use this to prevent all the device #defines from appearing in the list file.

Compatibility Notes

Preprocessor directives can be used to control various compile options. For example, the optimization level. The C standard recognizes that many of these directives are specific to a particular compiler. In order to accommodate this the standard has a special syntax called a pragma. When a pragma is encountered by a compiler that the compiler does not understand it will simply issue a warning and continue compiling. For example, the CCS C compiler has a preprocessor directive to specify the compiler should be case sensitive. The standard C way to specify this is like this:

```
#pragma case
```

Any C compiler should compile this but only those who support the case pragma will do what is intended.

In this book the pragma is omitted since the CCS C compiler does not require it for CCS-specific directives. Therefore expect to see just:

```
#case
```

All of the remaining preprocessor directives in this chapter are pragmas that are not standard C but are supported by CCS C.

Nonstandard Pragmas

#warning

Works like **#error** except only a warning is issued and compilation continues.

#use delay

This directive is used to tell the compiler what kind of oscillator is being used in hardware. This allows the compiler to set the right fuses for the chip and it will generate several built-in functions to control time. You have already seen one of those functions: **delay_ms(time)**. The **#use delay** is in the e3.h file. Here are some examples:

```
#use delay(crystal=12mhz)              // external 12mhz crystal
#use delay(internal=8000000)           // internal clock set to 8mhz
#use delay(crystal=4mhz,clock=16mhz)   // external 4mhz crystal and
                                       // multiply the speed by 4
```

The **#use** ... directives in the CCS C compiler cause the compiler to generate new built-in functions on the fly at compile time, according to programmer-specified options. There are **#use** libraries available for RS-232, I2C, touch panels, and many more. Details on these functions will be covered later in this book.

More traditional C compilers for larger machines will instead have C code libraries for common functions like **delay_ms**. Because of the tight memory requirements on the PIC® MCU processors, it is more effective to generate these functions at compile time so they will only include exactly what the programmer needs.

This directive is important for all programs because sometimes the compiler needs to insert delays to meet specific device interface requirements. For example, there may be PIC® registers that require a delay between setting and reading a bit. The compiler uses the clock information in this directive to know how to implement those delays.

Some chips have dual oscillators and other fancy clock features. This directive is designed to deal with all those options in a constant fashion across all devices.

About Frequency

Frequency is expressed in hertz, abbreviated Hz. Something with a frequency of 1 Hz is something that happens once a second; 2 Hz indicates it fully repeats and happens twice a second. The following are the common units used:

Hz hertz	1 per second
kHz kilohertz	1000 per second
MHz megahertz	1,000,000 per second
GHz gigahertz	1,000,000,000 per second.

#use rs232 (options)

This directive creates a set of functions to receive and transmit serial asynchronous data. There are many options; however, a simple example looks like this:

```
#use rs232( baud=9600, xmit=PIN_C6, rcv=PIN_C7 )
```

This creates functions such as **putc()** and **getc()** to transmit and receive characters over pins C6 and C7 at 9600 baud. If the pins are connected to a UART pin on the PIC® then the hardware is used. Otherwise the compiler creates a function to bit-bang the data. Asynchronous communication and the #use rs232 directive are covered in detail in Chapter 22.

#fuses options

options vary depending on the device and are listed in the device .h file for the device of interest.

The PIC® microcontrollers have an area of non-volatile memory that contains configuration settings. These are sometimes called configuration bits or fuses. These set things such as the oscillator type, enable/disable various hardware features, and sometimes set what various pins are used for. These settings are made when a program is loaded into the chip and do not change until the program is reloaded. The available settings vary between chips. To see all the settings available for your chip in the IDE use VIEW>FUSES. The CCS C compiler will set all the needed oscillator fuses as a result of the **#use delay** directive. Two other common fuses that apply to all chips are:

```
#fuses PROTECT
```

This prevents a program from being read from the chip memory once the chip is programmed. This is typically used during a products, production to prevent the end customer from making a copy of the program in the chip.

```
#fuses WDT
```

This enables a chip feature called the watchdog timer. The way it works is a timer runs until it reaches a preset value. Once that time is reached the chip is reset. Under normal operation the program will periodically reset this timer back to the start value. As long as the timer keeps getting reset the program keeps running. If the program gets stuck somewhere, the watchdog timer will restart the program after a set period of time passes. The **setup_wdt()** function is used to set the timeout time. The **restart_wdt()** function is used to reset the timer.

#locate id=address

id is a valid C identifier
address is a location in PIC® microcontroller data memory

Normally when a C variable is declared, the compiler finds the next free RAM location for the variable. This directive is used to force a variable to be located at a specific memory location. This can be used to force a C variable on top of a PIC® special function register. For example, on the PIC® used on the E3 board the status register is at RAM location 4056. The following code creates a C variable at that location:

```
int status;
#locate status=4056
```

After this a C statement like the following would set the status register to zero:

```
status=0;
```

The **getenv()** macro can be used in the CCS C compiler to obtain the location of a PIC® register. The following will do the same thing as the code above and it will work on all chips:

```
#locate status=getenv("SFR:STATUS")
```

Although there is no ANSI standard way to make a declaration like this in C, there is an embedded extension to C to do this. The extension to the standard for embedded programs is called named registers. The CCS C compiler supports this; however, very few compilers do support it. The same declaration using named registers looks like this:

```
register_status int status;
```

The underscore followed by a name looks up the register name address.

#byte id=x #word id=x

id is a valid C identifier
x is a C variable or constant

These are shortcut directives that work like **#locate**. In addition to locating the variable, it also declares the C variable. For **#byte** the type is 8 bits and for **#word** the type is 16 bits. The following does the same thing as the example in **#locate**:

```
#byte status=4056
```

#bit id=x.y

id is a valid C identifier
x is a C variable or constant
y is a constant 0–7

This is like **#byte** except the variable created is a single bit. A memory location is specified as well as the bit number in the byte that the C variable maps to. Bit 0 is the least significant bit in the byte. For example:

```
#bit carry_flag=4056.0
```

#reserve address

This directive reserves a RAM location so the compiler will not use it for anything. This may be helpful if there is another program in the same chip such as a bootloader.

Bootloaders

A bootloader is used as a separate program in the program memory that executes when a new application needs to be reloaded into the rest of program memory. The bootloader will use a serial port, USB port, or some other means to load the application. Frequently a bootloader will always execute on restart to check if a new program is to be loaded or if the application is to be run. Sometimes a bootloader will have primitive functions that the application also calls. This is the case for the E3 board. The bootloader is used to reload application programs and contain the USB functions so the application programs can transfer data with the PC.

#rom address={data}

This directive inserts data into the chip memory at the specified address. The most common use is to initialize the data EEPROM in the chip to some value. Many of the PIC® microcontrollers have internal data memory that will retain data even when the chip is powered down. The compiler has functions to read and write to that memory; however, sometimes there is a need to start the data out at some value. The address in the chip memory is different for each chip. To find the address there is a predefined macro that gets replaced by the preprocessor with the requested item. There are many things that **getenv()** can return. In this example we ask for the start of the EEPROM data memory:

```
#rom getenv("EEPROM_DATA")={0,0,0,0,123,255,255}
```

On the PIC16F887 chip this is the same as:

```
#rom 0x2100={0,0,0,0,123,255,255}
```

Another neat feature of **#rom** is it can calculate a checksum of all the program memory and store that checksum somewhere. One use of this is that a program could periodically calculate its own checksum and compare it to this value to make sure memory has not been corrupted. If we wanted to put the checksum at the last memory location it would look like this:

```
#rom getenv("PROGRAM_MEMORY")-1=checksum
```

The −1 is needed because addresses start at 0. In a chip with 1024 words of memory, they are in locations 0–1023.

The checksum is a simple sum of the what is in every memory location except the location that the checksum is inserted into.

Another common use for **#rom** is to insert data at a specific address because it will be changed by some external means. For example, encryption codes that are loaded in during production depending on the specific customer. Another example would be specific device configuration data, like number of ports on the product. The **#rom** data can be read by the program with a function call like **read_program_memory()**, where the absolute program memory address is used to read the data.

#id data

Many chips have a special area of memory (like one word) where the programmer can put some kind of identification. This area can be read even when the program memory is protected from reading. It can be used to save a firmware revision number or as another place to put a checksum like this:

```
#id checksum
```

Other Pragmas

The following list is of other common preprocessor directives. For a full list use the compiler reference manual under the preprocessor section.

#device	Used to tell the compiler what specific chip the program is to be loaded onto.
#hexcomment	Inserts a comment into the final output file; may be used to identify a version or configuration.
#zero_ram	Tells the compiler to zero out all RAM locations before the program runs. This may help to make programs run more consistently if there are variables used before being initialized to a value.

#fill_rom Causes all unused program memory locations to have some preset value. This is done if a programmer is concerned that a program that loses control might start executing in an area that could cause damage.

#pin_select Some chips allow a given pin to be attached to one of many modules. This directive allows the programmer to select what the pin should be attached to. On chips that have reprogrammable pins it is very important to use this directive before the peripheral is referenced in the code. For example, on many PIC24 parts the internal UART pins are reprogrammable. Be sure to have #pin_select directives before the **#use rs232**. For example:

```
#pin_select U1RX=PIN_B0 // assign B0 to UART 1 receive
#pin_select U1TX=PIN_B1 // assign B1 to UART 1 transmit
```

#serialize This is used to insert instructions in the output file for the device programmer to put a unique serial number in each chip that is programmed.

#inline This directive put before a function causes that function to be copied to wherever it is called instead of calling the function.

#separate Opposite of inline; tells the compiler to not inline this function. By default the compiler decides if it is best to call or inline.

#int_... Another group of function qualifiers, this marks a function as an interrupt function. Interrupts are covered in more detail in Chapter 17.

Summary

- Preprocessor directives are processed after tri-graphs are processed and comments are removed, and before all other C code.
- Preprocessor identifiers are separate from regular C identifiers.
- Preprocessor replacements are done on a textual basis.
- Preprocessor directives can be used to improve the readability of code.
- Preprocessor directives can make it easy to change key program characteristics.
- Preprocessor directives can be used to maintain one code base for multiple program configurations.
- Preprocessor directives can be used to access compiler- or processor- specific features.

Exercise 3-1

Objective: Become familiar with the use of the basic preprocessor directives **#define** and **#include**.
Requires: E3 module, USB cable, PC.

Steps/Technical Procedure	Notes
1. Write a program that turns the green LED on for 10 seconds, the yellow LED on for 3 seconds, and the red LED on for 10 seconds. There is a special form of the **for** statement that loops forever. The syntax is for(;;;) statement. Use **#define**s for the pins and the times so they can be easily changed. Compile, load, and test the program.	
2. Make a X2-3.c program that adds a **#define** to the top of the file to define TYPE. TYPE can be 1, 2, or 3. When it is 1 the program should have the times 10, 3, 10, as above. Type 2 should use 10, 3, 5 seconds, and type 3 should be 20, 3, 5 seconds. Conditional compilation directives should be used to deal with TYPE. Compile, load, and test three times, changing only the one line of code.	
3. Make a X3-3.c program where the main loop has only three lines in it. Each line should be a define reference like: HANDLE_GREEN_LIGHT. The **#define**s should turn the LED on, delay, and turn it off. Compile, load, and test.	
4. Make a X3-4.c program like X3-3.c but instead of three define lines in the main loop put in three **#include** lines. Each include file should have the needed statements to deal with each of the three lights. There will be three include files: X3-4-1.c, X3-4-3.c, X3-4-3.c. Compile, load, and test.	
5. Make a X3-5.c program that uses only one include file (X3-5-1.c). In the main program, before each **#include** add two **#define**s to define the LED to use and the on time. Those defines should be then used in the include file. Compile, load, and test. If you got warnings from your program, figure out how to eliminate them.	

6. Start with the X3-1.c program and make a X3-6.c program that works exactly the same but only has one ; character in the whole file. The only include file allowed is the e3.h. Compile, load, and test. Once it is working, destroy the program and don't do anything like that again.	
7. Find out what happens if you have a define like the following and then use it in your code: **`#define APPLE APPLE`**	

Quiz

(1) Given the following lines of code inside a function, what will happen?

```
int  x;
#define DELAY=5
x=DELAY;
```

(a) Program will delay for 5 seconds
(b) Variable x will be assigned 5
(c) Syntax error on line 2
(d) Syntax error on line 3
(e) Nothing will be done

(2) The trick used to make example X2-6.c such that a semicolon appeared only once in the file shows the power of the preprocessor. What syntactical item would this not work for?
(a) "
(b) < or >
(c) (or)
(d) #
(e) *

(3) What happens if a #define uses its own identifier name in the text of the define?
(a) This is the only way to get that identifier in the code post-preprocessor
(b) An error will be flagged on the #define line
(c) An error will be flagged where the define is used
(d) When the identifier is used it is turned into white space
(e) The computer hangs because it replaces the identifier with itself forever

(4) What happens in the CCS C compiler when you compile the following code?

```
1 #define X   5
2 void main(void) {
3 int x
4 x=X;
5 }
```

(a) The variable x is assigned 5
(b) Error on line 1
(c) Error on line 3
(d) Error on line 4
(e) No error, but the code does nothing

(5) Which line in the code below is invalid?

```
1 //////////////////////////////////////////////////////////////////////
2 main() {
3 {{{{{{{{{{{{{{{{{{{{{{{{{{{{{{
4 ((((((((((1))))))))))
5 ;;;;;;;;;;;;;;;;;;;;;;;;;;;;;;;;;;
6 }}}}}}}}}}}}}}}}}}}}}}}}}}}}}}
7 }
```

(a) Line 1
(b) Line 3 (and 6)
(c) Line 4
(d) Line 5
(e) There are no errors

(6) For the code in question 5, what is the first warning?
(a) Line 1, comment line missing text
(b) Line 2, missing void
(c) Line 3, code does nothing
(d) Line 3, duplicate semicolons
(e) None, there are no warnings that apply

(7) What warnings are output from the below code?

```
#define DEBUG
#ifdef  DEBUG
#warning one
#else
#ifdef  RELEASE
#warning two
#else
#warning three
#endif
#endif
#warning four
```

(a) One, three, four
(b) Four
(c) Two, four
(d) One, four
(e) One, three

(8) What is the variable name that results from the following?

```
1 #define APPLE   ORANGE
2 #define BANANA  APPLE
3 #define PEAR  BANANA
4 int  PEAR;
```

(a) None, this is an error at line 4
(b) PEAR
(c) BANANA
(d) APPLE
(e) ORANGE

(9) Given the following two include files and main program, what happens when an attempt is made to compile:

```
file1:
#ifndef  xx
#define xx
int  x;
#endif
```

```
file2:
```

(a) No errors
(b) Error in file 1, attempt to create a variable twice with the same name
(c) Error in main.c, the variable x is not known
(d) Error in main.c, the variable y is not known
(e) Error in main.c neither x nor y is known

(10) Given the following code, what line flags the first error?

```
1 #define  APPLE  5;
2 void main(void) {
3       int x;
4       x=APPLE;
5       x=APPLE/2;
6 }
```

(a) None, there are no errors
(b) Line 1
(c) Line 2
(d) Line 4
(e) Line 5

CHAPTER 4

Data Variables and Types

Data Types

All data in C, whether a constant, expression, or variable, has an associated type. That type describes the range, interpretation, and storage method of the data.

The most basic data types are character, integer, and floating point. Each of these has multiple variations and qualifiers that apply. Each will be covered in this chapter. Not covered until future chapters are the arrays, pointers, unions, and structures. Those types are made up of these three basic types.

C has specific rules for how and when it converts from one type to another when two different types appear in the same expression. This is covered in the next chapter.

Characters

The character was described in detail in the previous chapter. C characters are always a single byte. It should be mentioned that there is a trend in newer languages and language extensions to use two bytes for a character. This allows for specifying a character set (such as ASCII) as well as the character. The result is better compatibility with international alphabets. The C key word for a character is char and it is used like this to declare a variable:

```
char c;
```

Types can have qualifiers to further describe the type. In the case of a character there are two somewhat unnatural qualifiers that may be used:

```
unsigned char
signed char
```

Since C allows a character to appear anywhere a number appears, these numeric qualifiers are allowed. Stranger yet, most C compilers default a **char** to signed if it is not specified by the programmer. When using characters as characters (as opposed to numbers), the signed/

Embedded C Programming. http://dx.doi.org/10.1016/B978-0-12-801314-4.00004-1

unsigned does not matter. It will matter when characters are mixed in numeric expressions. For example:

```
signed char c;
c = '\xFF';
```

In this case **c** is not equal to **0xFF** because **c** as a signed 2's complement number is −1 and **0xFF** is 255.

Integers

The most commonly used type, and frequently most confusing, because C does not have a standard size for the integer types. Furthermore, it has multiple ways to specify the same type. The standard types are:

short	also known as **short int**
int	also known as **int int**
long	also known as **long int**
long long	also known as **long long int**

All the C standard says is the **int** should be the most natural size for the environment and the **short** is smaller and **long** is larger. A **long long** is not required to be implemented but frequently is. Like the **char**, each of these types can be qualified with **signed** or **unsigned**. In addition, the CCS C compiler has nonstandard names for each type. Table 4.1 attempts to clarify all this for Microchip 8-bit processors.

Compatibility Note

With the CCS C compiler a **#type** preprocessor directive may be used to set the actual number of bits in each standard C type. Table 4.1 represents the default.

Many C compilers specify the short type as 8 bits, and some programmers may use **short** and **char** interchangeably because they assume both are 8 bits.

Table 4.1 Integer types with ranges.

Standard Type	CCS C Alias	Bits in Memory	Signed Range	Unsigned Range
short	int1	1	N/A	0–1
int	int8	8	−128 to 127	0–255
long	int16	16	−32,768 to 32,767	0–65,535
long long	int32	32	−2,147,483,648 to 2,147,483,647	0–4,294,967,296

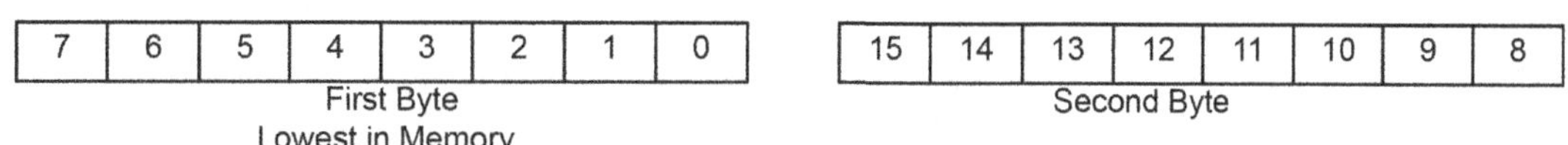

Figure 4.1: 16-bit integer in memory.

Most C compilers default types to signed where in the CCS compiler the default qualifier is unsigned. This also can be changed with **#type**.

Where for 8-bit chips the CCS C compiler defaults to unsigned and bit sizes 1, 8, 16, and 32, for the 16-bit chips (PIC24 and up) the defaults are signed and 8, 16, 32, and 64. That compiler also has additional integer data types of **int48** and **int64**.

The standard CCS header files also define types of BYTE (same as **int8**) and BOOLEAN (same as **int1**).

Integer Format

When dealing with a multi-byte integer, there are two ways to store the integer in memory. The C standard does not dictate a specific method. Most PIC® C compilers use what is called a little endian format. This means the low byte (least significant) is saved in the lowest (first) memory location. Big endian formats store the bytes in the opposite order. For example, a 16-bit integer is saved in little endian format as shown in Figure 4.1.

Enumerated Types

C has a special way to create an integer subtype that allows the programmer to give unique names to each number in the type. Consider this example:

```
enum {RED, GREEN, BLUE, YELLOW, BROWN} color;
color=BLUE;
```

The **enum** line creates a new subtype. The intended values for the variable color are the listed identifiers. The assignment line assigns BLUE to color. In reality 2 is assigned to color. By default the compiler treats RED as 0, GREEN as 1, and so on. If the programmer wants to use specific numbers it may be done so like this:

```
enum {RED=10, GREEN=20, BLUE=30, YELLOW, BROWN} color;
```

In this case because YELLOW and BROWN were not specified they will be 31 and 32 (following after the BLUE value).

enum has another form that allows the reuse of the type by creating what is called a tag. The tag is specified after the **enum** key word and when used the variable name is optional:

```
enum colors {red, green, blue, yellow, brown };
enum colors forground_color;
enum colors background_color;
```

Fixed Point

Fixed point is implemented as a qualifier to an integer type. To specify a long integer with two decimal points fixed, do this:

```
long _fixed(2) x;
```

In this case the qualifier comes after the basic type. Notice the underscore before the fixed key word. Here are some operations:

```
x = 1.23;        // Actually assigns 123 to x
x = 5;           // Actually assigns 500 to x
```

In our example the range of x is 0–655.35.

Floating Point

There are two defined floating-point types in standard C: float and double.

There is no standard for the range and precision of a floating-point variable. Table 4.2 details the CCS C implementation. For 8-bit processors only the 32-bit float is supported.

On microcontrollers, floating-point operations take a lot of time and space in memory. Whenever possible integer or fixed-point math should be used.

Interpretation Help

It should be clear now that the same data in memory can mean different things depending on the type. The IDE has a tool to help working with these formats. Use Tools>Base Converter to start the utility. For example, enter a 123 in the unsigned field and see 7B in the Hex field.

Table 4.2 Floating-point types, ranges, and accuracy.

Standard Name	CCS C Alias	Bits in Memory	Signed Range	Digits of Precision
float	float32	32	-1.5×10^{45} to 3.4×10^{38}	7-8
double	float64	64	-5.0×10^{324} to 1.7×10^{308}	15-16
	float48	48	-2.9×10^{39} to 1.7×10^{38}	11-12

Floating-Point Format

The CCS C compiler uses the original Microchip format in memory for floating-point numbers, as shown in Figure 4.2. Note that the mantissa is not saved in 2's complement form (so it needs a sign bit). The exponent is in 2's complement form; however, it has an offset added, called a bias.

Another format that is popular for floating-point numbers on a PC is the IEEE standard format. It looks like Figure 4.3.

There is an implied 1 in the nonexistent 24th bit of the mantissa. For example:

24-bit mantissa with decimal indication = 11.00000000000000

To move the decimal point to the normalized location (after the imaginary 1) an exponent of 1 is needed, adding the bias of 7F you get 80 hex. The hex representation is then:

80 40 00 00

3.5 would be:

80 60 00 00

7.0 would be:

81 60 00 00

Again the Base Converter in the IDE is helpful to see these relationships.

Void

Void is both a key word and type placeholder. It can appear in many places that a type can appear. It means essentially no type. It may be used to indicate there is no specific type. Recall in Chapter 1 it was used to indicate **main()** did not have any parameters and did not return anything. It may not be used to declare a variable like this:

```
void x;
```

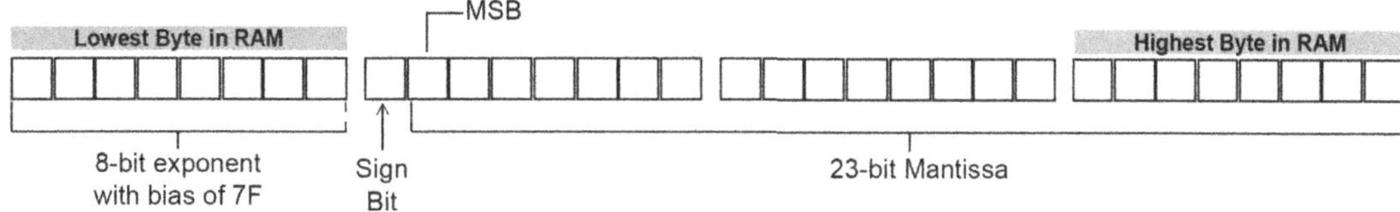

Figure 4.2: Traditional Microchip floating-point format.

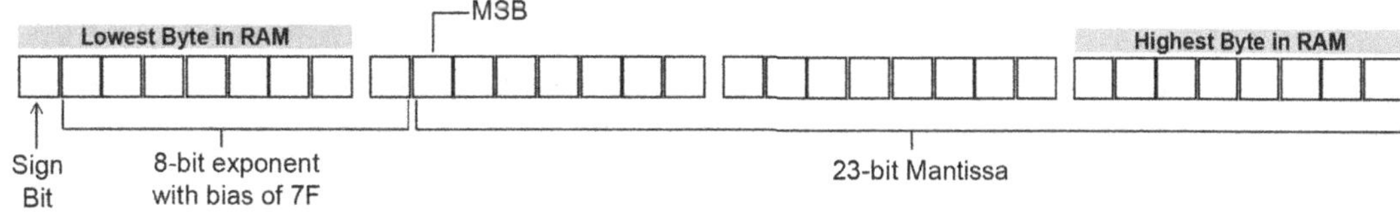

Figure 4.3: Standard IEEE 32-bit floating-point format.

In this case the type needs to be fully known because space must be allocated. Void will be covered again when discussing pointers in Chapter 10.

typedef

C has a mechanism to create a new name for a type. Once a new name is created it can be used anywhere a standard type (like int) is used. The basic format is:

```
typedef old-type new-name;
```

Here are some examples:

```
typedef   long  WORD;
typedef  unsigned long long  UINT32;
typedef   enum  { RED,  GREEN, BLUE,  YELLOW, BROWN }  COLORS;


WORD  x;
UNIT32  y;
COLORS  z;
```

That last one is a bit strange. What happens here is a type is created named COLORS (uppercase typedefs is a style thing), so COLORS may be used like we were using **int**. For example:

```
COLORS  color;
color = yellow;
```

What we did here was eliminate the need to type **enum** to declare new variables.

Declaring Variables

Variables take some space in the processor memory (usually RAM). A simplified declaration format for variables looks like this:

optional-qualifiers type identifier-list ;

The optional qualifiers are key words like **static** and **const**. The type is a key word like **char** or **int**, or a user-defined type (via **typedef**). The identifier list is a list of variable names separated by commas. Optionally each identifier can be followed by an initializer; this is an = and a constant value. Some examples:

```
int x;
int x,y,z;
unsigned char c = 'A';
```

Static will be covered in more detail in the section Life of a Variable. The initializer is used to set the variable to an initial value.

Identifiers

Names for constants and variables are referred to as identifiers (id). The following are specific rules:

1. All names used in the same scope must be unique.
2. Names must begin with a letter of the alphabet or an underscore.
3. After the first letter, names can be made up of letters, numbers, and underscores in any combination. An occasional capital letter may be used for clarity.
4. Names may be of any length; however only the first 32 characters are used by the compiler.
5. ANSI (standard) C is case sensitive, but CCS C is not by default.
6. Identifiers that start with one or two underscores followed by an uppercase letter are considered to be reserved for compiler use by the C standard. No compiler enforces this and most programmers ignore it.
7. Table 4.3 lists key words that are reserved for special uses in C. They cannot be used.

In addition, Table 4.4 lists nonstandard key words reserved by the CCS C compiler.

Table 4.3 ANSI C key words.

auto	enum	signed
break	extern	sizeof
case	float	static
char	for	struct
const	goto	switch
continue	if	typedef
default	int	union
do	long	unsigned
double	register	void

Table 4.4 CCS C additional key words.

addressmod	float64	int32
_fixed	int1	in48
float32	int8	int64
float48	int16	

Scope of a Variable

Variables declared outside a function are said to have global scope. Any function after the variable is declared may access this variable. Variables declared inside a function are local variables. Local variables may only be accessed inside the function they are declared (again, after the declaration). There is a third level of scope in C. If a variable is declared inside a { } block (compound statement) then it can only be accessed inside that block. There is a special extension to that rule that allows a variable to be declared inside the ( ) of a for statement and the variable may be accessed throughout the whole for loop. Consider this program structure:

```
1: int  ga;
2:
3: void f1(void) {
4:    int la;
5:    {
6:       int lb;
7:
8:    }
9: }
10:
11: int gb;
12:
13: void f2(void) {
14:  int lc=1;
15:
16:    for( int ld = 5; i<=10; i++ ) {
17:
18:    }
19: }
```

Variable ga can be accessed from lines 1 to 19.
Variable gb can be accessed from lines 11 to 19.
Variable la can be accessed from lines 4 to 9.

Variable lb can be accessed from lines 6 to 8.
Variable lc can be accessed from lines 14 to 19.
Variable ld can be accessed from lines 16 to 18.

It is allowed to give the la variable instead of the name ga. Doing so would make the global ga inaccessible from inside function f1. Global variables are always in memory. If static is applied to a global variable then that variable is initialized to 0 like other static variables. These variables are called external static and the variables cannot be accessed outside this compilation unit. Since our programs so far have been a single compilation unit, the only effect of a global static or external static is the 0 initialization.

Sometimes a company coding standard will dictate that global variables identifiers have a special prefix (like global_, g_, G, or g) in order to recognize them as global variables. Sometimes this will also include the module or file nickname to help find them.

In addition to variables having scope, typedefs also have the same kind of scope.

Life of a Variable

In addition to scope, variables have a life-span. By life, we mean the time the variable is taking up memory. Global variables are always in existence. Common local variables, such as previously listed, are destroyed when the scope terminates. For example, if function `f1` is not currently executing then `la` is not in memory. More importantly, if `f1` is executing and assigns a value to `la`, then when the function is called again that value is not still there. There is a `static` type qualifier, however, that will cause a local variable to never be destroyed. It is used like this:

```
static int la;
```

In this case when `f1` is called `la` starts with the same value it had when `f1` last ended.

Notice in `f2` we initialized `lc` to 1. This happens each time `f2` is called. If we would have declared `lc` static then the 1 would only be assigned on the first call to `f2`.

One additional side effect of static is that all static variables are initialized to 0 when the program starts unless you initialize them to something with an initializer.

The only reason to declare a global variable static is to initialize it to 0. It is, however, less typing (and clearer) to just add a `=0` after the declaration.

More Qualifiers

The standard C qualifiers are listed in Table 4.5.

CCS C compilers also have the qualifiers shown in Table 4.6.

Table 4.5 Standard C type qualifiers.

static	For variables, initializes them to 0 and prevents the variable from being destroyed
register	Tells the compiler to use the most efficient memory for the variable
signed	For integer and character variables the encoding is 2's complement
unsigned	For integer and character variables there is no sign bit
volatile	Indicates the variable might change outside the normal program sequence Used for variables that may be changed by interrupt functions or hardware
const	Variable is read-only implementation-dependent coding
extern	No storage is to allocated for the variable. Another declaration without extern will appear elsewhere in the code. It is there that storage is allocated
auto	Not commonly used. Was intended to be the opposite of static

Table 4.6 CCS C additional type qualifiers.

_readonly	True read-only qualifier without changing memory type
_fixed(n)	Specifies a fixed-point type
rom	Indicated variable is to located in program memory
_packed	For structure types, indicates no compiler-generated unused space is to be added

Design Documentation

It is a good practice to include a comment after each global variable to indicate how the variable is used. When formal documentation for a program is required a document will contain all the global variables, and how they are initialized and used within the code. The specific requirements may be specified in a company coding standard. There are tools such as the Document Generator in the CCS IDE that can extract the comments from the code and produce at least a first draft of a data definition document.

Comments just before a variable declaration or after it on the same line will be associated with the variable. The following are some examples:

```
int8  start_down_time;   // Number of cycles the start button was
                         // held down
int8  start_up_time;     // Number of cycles the start button is
                         // not down (max 99)
int1  start_pressed;     // TRUE is debounced buton has been pressed

                         // and not processed line_voltage is calculated

                         // by taking the peak voltage into AN5 (from D9)

                         // and multipling by 6 (for the resistor divider)
                         // and by 10 (for the transformer)

                         // to get the AC peak voltage for the last half
                         // second.
float line_voltage;
```

RAM

Variables are saved in random access memory (RAM). RAM data is lost after power cycle. Photos showing advancements in physical RAM are given in Figures 4.4–4.6).

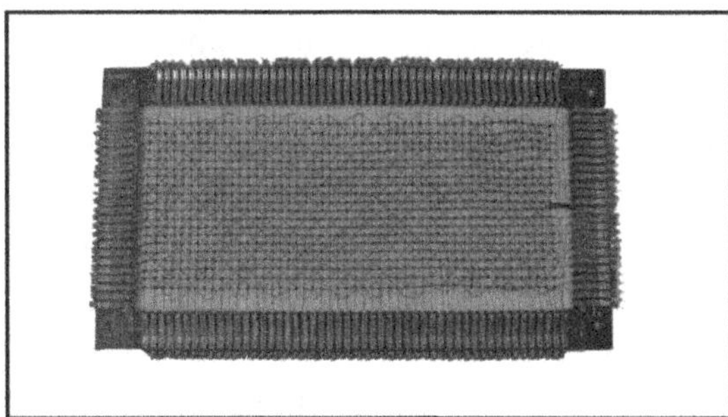

Figure 4.4: 50 x 20-bit core memory module from an early computer. Total of 1000 bits.

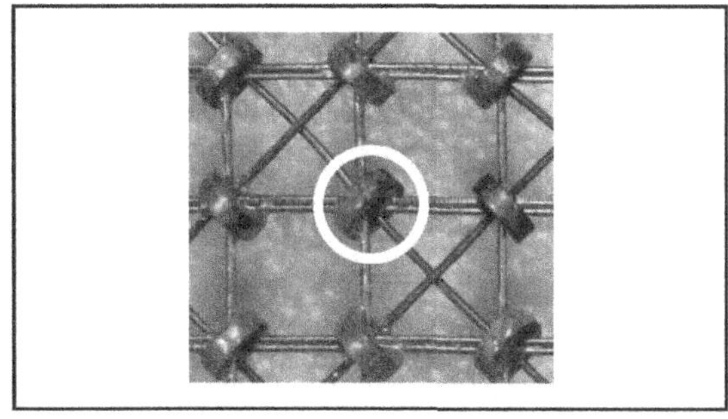

Figure 4.5: Single bit of the Figure 4.4 memory core. Each bit is magnetized to indicate 0 or 1.

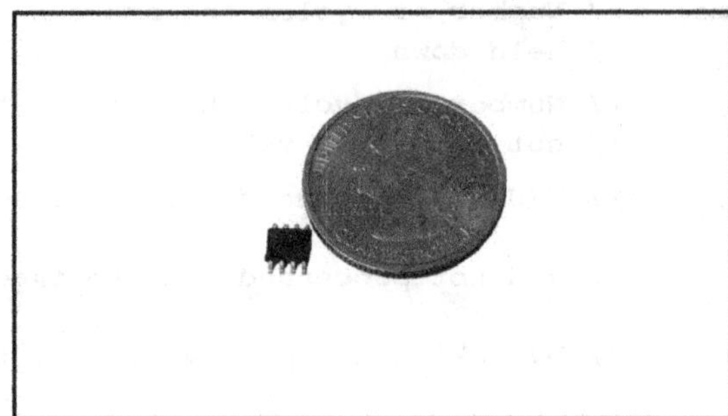

Figure 4.6: Modern 256K x 8-bit memory chip. Total of 2,097,152 bits.

Computer memory is usually measured in bytes. "K" is used for 1024 bytes (not 1000 as in electronics) and "M" for 1024*1024 bytes. "G" is for 1024*1024*1024 bytes.

Summary

- All data, both variables and constants, has an associated type that indicates the machine encoding and interpretation.
- The basic types are character, integer, and floating point.
- The variations in each basic type include types that specify the number of bits. This will indicate a specific range for the data type.
- Data types have qualifiers that may be used to indicate specific characteristics of the type.
- Enumerated types allow for giving names to numbers in a named group.
- **`typedef`** is used to create a user-defined type that is a variation of existing types.
- Variable identifiers start with A–Z or underscore and must follow specific rules.
- Variables have scope that determines where in the code they can be accessed.
- Variables have a life-span that determines when the variable exists and when that memory location may be reused.

Exercise 4-1

Objective: Write a program to determine how many bits a variable that is a simple **enum** with a few elements takes.
Requires: E3 module, USB cable, PC.

Steps/Technical Procedure	Notes
1. Write and compile a simple program with a main and no variable declarations. Use the C/ASM viewer to check the top of the LST file and see how much RAM is being used. This is the RAM the compiler and E3 interface code are using. Write this number down. Add a simple enum with six items. Recompile and check the RAM used. Subtract the RAM used without the **enum** to determine how many bytes were added for the new variable. From this number determine how many elements could be in the enum with this storage size.	
2. Find out if the limit determined in X4-1 is an absolute limit of the compiler of if the storage size changes with the number of elements. Change the **enum** so the first element has **a=n** after it, where n is the maximum number determined in X4-1 minus 2. This will cause the numeric values for the last three elements to be over the storage ability. Compile and check the LST file for the RAM usage.	
3. Another file that will indicate the specific RAM storage for every variable is the .SYM file. Click on SYMBOL MAP on the compile ribbon to view this file. Check the actual RAM locations and number of bytes for X4-1 and X4-2.	
4. C has a very handy built-in function to find out how many bytes a variable or type takes. It is **sizeof(e)**, where e is any type, variable, or expression. In the case of an expression, **sizeof** returns the space for the type of the expression. With a **printf** statement, use **sizeof** to confirm the results of X4-1 and X4-2.	
5. The formal syntax for C allows for some strange-looking types. The following are some examples. Use **sizeof** to figure out how much space is allocated for each of these odd-looking types: 1. **int int int int** 2. **long long int int** 3. **long int long**	
6. Declare four variables of type int1 and find out how much storage space (via the SYM file) is used for the four variables.	

Quiz

(1) A program takes three samples that each has a value of 0 to 100. It sums those samples and divides by 3 to get an average. What is the best data type for the sum?
   (a) short (or int1)
   (b) int (or int8)
   (c) long (or int16)
   (d) long long (or int32)
   (e) float

(2) For the same problem as in question 1, what would the maximum sample value be that this type could accommodate?
   (a) 255
   (b) 65,535
   (c) 100
   (d) 21,845
   (e) 1023

(3) Which of the following declarations are in error?
   (a) **`enum { short, tall, giant } size;`**
   (b) **`int a=1, const b=2, static c=3;`**
   (c) **`static void x;`**
   (d) b and c are invalid
   (e) All are invalid

(4) Which of the following are not valid identifiers for a local variable in a function called by main()?
   (a) _123456
   (b) GBCDEFAHIJKLMNOPQRZWXYZ123456
   (c) for
   (d) main
   (e) None are valid

(5) What is the tag on the following typedef?

```
typedef enum ccc {purple, magenta=4, fuchsia}
ddd;
```

   (a) ccc
   (b) 4
   (c) fuchsia
   (d) ddd
   (e) typedefs don't have tags

(6) How much storage space does a variable declared as a enum take?
 (a) The maximum size int that the compiler supports
 (b) The smallest size int that the compiler supports
 (c) The smallest size int that will fit the largest value in the list
 (d) An enum variable does not take up RAM
 (e) Depends on a qualifier to specify the storage space

(7) If you want to use the identifiers TRUE and FALSE in a program using a compiler that does not have them defined in a header file, what method may be used to do that?
 (a) **`#define`**
 (b) **`const int`**
 (c) **`enum`**
 (d) a or b
 (e) a, b, or c

(8) One might consider using a #define instead of a typedef to create a new type (like WORD). Which of the following statements is true concerning this?
 (a) **`#define`** won't work, that is why we have a **`typedef`**
 (b) There is no difference between the two methods
 (c) **`#define`**s don't work inside functions
 (d) **`typedef`**s have a limited scope when used in a function
 (e) All are true

(9) Which of the following declarations is in error?
 (a) **`enum {a='A',b='B',c='C'} letters;`**
 (b) **`const int MAX;`**
 (c) **`typedef int integer;`**
 (d) **`float threshold=1;`**
 (e) **`signed char c=-1;`**

(10) When should a floating-point variable be declared?
 (a) When the code must run very fast
 (b) Whenever the application requires math with a decimal point
 (c) When space in the chip is very tight
 (d) When the accuracy of the numbers must be exact
 (e) When the range of numbers is beyond the range of the largest integer

CHAPTER 5

Expressions and Operators

Mathematical Operators

The most useful mathematical operators are (no surprise):

- `+` addition
- `-` subtraction
- `*` multiplication
- `/` division

We will use them in examples. You may also encounter the related:

`%` modulus

Modulus produces the remainder from division. For example:

15 % 6 evaluates to 3. 15 divided by 6 yields a remainder of 3.

To avoid confusion, note that the symbol `%` is also used to format variables in a string using the **`printf()`** built-in function.

Some operators use the same symbol for two different operations. The above are binary operators. The general form is:

operand operator operand

There are two operands (or expressions) for binary operators. The `-`, for example, can also be a unary operator, like this:

```
-x
```

In this case the `-` operator will return the expression with a changed sign, in this case x. There is also a `+` unary operator; however, it does not do anything but return the operand.

Compatibility Notes

Some math operations cause a result that cannot be represented by the specified type. The standard allows the compiler to decide how to handle this. For example, a byte can have 0–255. Some compilers might choose to treat 255 + 1 as an error (overflow). Since error

Embedded C Programming. http://dx.doi.org/10.1016/B978-0-12-801314-4.00005-3

messages are not practical in an embedded system, most compilers will simply do modular arithmetic. 255 + 1 is 0 and 255 + 2 is 1, and so on.

The results are less predictable with floating-point numbers. Sometimes the number pegs at the maximum value and sometimes you get garbage.

A divide by 0 is also unpredictable. If any of these conditions are important to your program operation you should have specific tests for them before the operation is performed.

Operator Precedence

Operator precedence is important. Rules determine which mathematical operation takes place first, i.e. takes precedence over others. Parentheses, `( )`, may be used to force an expression to a higher precedence. See Table 5.1.

Operators higher in the chart have a higher precedence. `()` is higher than `+`.
1 has higher precedence than 2 which has higher precedence than 3.
A sub-expression in parentheses is evaluated first, regardless of the operators involved.
If there are two or more operators having the same precedence, they are usually evaluated left to right; some operators have a right to left rule and those will be identifier later.
Knowing the precedence rules is good; however, it is better to add `()` to your expressions when it is important to improve readability.

Examples:

2+3 * 4	evaluates to 14 (multiplication first, addition second).
(2+3) * 4	evaluates to 20 (inside parentheses first, multiplication second).
(2 * (4+(6/2)))	evaluates to 14 (inside parentheses first, work outward).
4 * 5/2 * 5	evaluates to 50 (start left, move right).

Table 5.1 lists all the operators, their precedence, and the direction they associate (left to right or right to left).

Expression Type and Type Conversion

In addition to the order of operator evaluation, it is important to understand the expression type. The result of every expression has a type, just like a variable has a type. If both operands

Table 5.1 Precedence.

Precedence	
()	1
* / %	2
+ -	3

of a binary operation are the same type, then the expression type is that type. This might seem obvious until you consider something like this:

```
int8 x;
int8 y;
int16 z;
x=10;
y=30;
z=x*y;
```

In this case **z** will not be 300 because 300 is larger than a byte and the result of **x*y** is a byte. **z** is actually 44. To find out why you need to picture the 300 in memory, it is 0x12C. Since a int8 has only two hex digits the actual result will be 0x2C. 0x2C is 44 decimal.

If the two operands are of different types then there is a type conversion made according to rules of C called "usual arithmetic conversion." The simple way to understand usual arithmetic conversion is the smaller type operand is converted to the larger type operand. For example, if one operand is an int8 and the other is an int16 then the int8 operand is converted to an int16 before the operation is performed. Note, in the case where the operand is a variable, the actual variable value does not change, only the value used for the operation.

When C does a type conversion, every attempt is made to represent the same value in the new type. However, this is not always possible. For example, a conversion from 1.23 to an integer type will result in a 1 since the fraction cannot be represented. In the above example, the assignment operator performed an automatic type conversion from the int8 result (of multiply) to the lvalue variable **z** (an int16). This conversion happened after the number was truncated to a byte.

The C programmer can also force a type conversion. This is called a type cast. The above example can be fixed with any of the following type casts:

```
z=(int16)x*y;
z=x*(int16)y;
z=(int16)x*(int16)y;
```

A type cast is a type name inside parens before an expression. The above works because the type cast has a very high precedence. The single type cast works because the usual arithmetic conversion rules cause the other operand to be automatically converted to a int16 when the other one was type cast. The following would not work:

```
z=(int16) (x*y);
```

because the multiply would be done before the type cast.

In addition to binary operands that are not similar, including the binary assignment operator, C also does an automatic type conversion for function arguments.

Relational Operators

Relational operators all do some kind of evaluation and then return a 1 (for true) or 0 (for false). Table 5.2 shows the operators.

To help understand how these operators are used together, consider these examples:

<table>
<tr><td><code>((a==1) || (a==2) || (a==3))</code></td><td>True if a is a 1,2 or 3</td></tr>
<tr><td><code>((a>='A') && (a<='Z'))</code></td><td>True if a is a capital letter</td></tr>
<tr><td><code>(!debug_mode && (a>99))</code></td><td>True if debug_mode is 0 and a is over 99</td></tr>
</table>

The relational operators are evaluated left to right. In some cases the remaining operations in an expression become irrelevant. In this case they are not evaluated. This is called a short circuit. For example:

```
( (a==1) || (a==2) || (a==3) )
```

If **a** is equal to 1 then the **a==2** and **a==3** comparisons are not done. This is important to know if an expression has a side effect. For example, say instead of **a==3** we had **find_mode()==3**. In this case there is a function call in the third term. That function returns a value but it might also do other things like keep a count of calls or output data to the screen. When **a** is 1, that function will not be called due to the short circuit rule.

Table 5.2 Relational operator list.

<table>
<tr><td>==</td><td>Equal to</td><td>a==b returns 1 if a and b are the same</td></tr>
<tr><td>></td><td>Greater than</td><td>a>b returns 1 if a is larger than b</td></tr>
<tr><td><</td><td>Less than</td><td>a<b returns 1 if a is smaller than b</td></tr>
<tr><td>>=</td><td>Greater than or equal to</td><td>a>=b returns 1 if a is larger than or equal to b</td></tr>
<tr><td><=</td><td>Less than or equal to</td><td>a<=b returns 1 if a is smaller than or equal to b</td></tr>
<tr><td>!=</td><td>Not equal to</td><td>a!=b returns 1 if a and b not the same</td></tr>
<tr><td>&&</td><td>And</td><td>a&&b returns 1 if a is nonzero and b is nonzero</td></tr>
<tr><td>||</td><td>Or</td><td>a||b returns 1 if a is nonzero or b is nonzero</td></tr>
<tr><td>!</td><td>Not (unary)</td><td>!a returns 1 if a is zero</td></tr>
</table>

Table 5.3 Binary bitwise operator list.

<<	Left shift	0b00000101 << 2 will yield 0b00010100
>>	Right shift	0b00001100 >> 2 will yield 0b00000011
&	Bitwise AND	0b00001111 & 0b11111000 will yield 0b00001000
|	Bitwise inclusive OR	0b00001111 & 0b11111000 will yield 0b11111111
^	Bitwise exclusive OR	0b00001111 & 0b11111000 will yield 0b11110111

Binary Bitwise Operators

Binary operators (or bitwise operators) are mathematical operators that view the data from a binary perspective. These operators are not permitted on float types. See Table 5.3.

The AND and ORs all do an operation on a single bit of the first operand with the corresponding bit of the second operand to get a bit for the result. This is done for every bit in the data. The truth tables for how the bit-by-bit comparison is done are shown in Tables 5.4–5.6.

Table 5.4 AND & bit truth table.

AND &		
1st Operand	**2nd Operand**	**Result**
0	0	0
1	0	0
0	1	0
1	1	1

Table 5.5 OR \| bit truth table.

OR |		
1st Operand	**2nd Operand**	**Result**
0	0	0
1	0	1
0	1	1
1	1	1

Table 5.6 Exclusive OR ^ bit truth table.

Exclusive OR ^		
1st Operand	**2nd Operand**	**Result**
0	0	0
1	0	1
0	1	1
1	1	0

Compatibility Notes

The shift operators shift the number on a binary basis. If the shift causes bits to go outside the range of the data then those bits are lost For example **3>>1** is a 1. Note that shifting a signed number can produce unpredictable results. Let's look at an example of a signed byte representing −4:

11111100

If we shift this data to the right by one bit, then there are two ways the compiler might do this operation:

1. 01111110 yielding a 126 decimal
2. 11111110 yielding a −2

The second method is called an arithmetic shift. In an arithmetic shift the sign of the number is not affected. The C standard allows the compiler to do either a logical shift or an arithmetic shift for signed data.

Assignment Operators

We have already been using the primary assignment operator = in many of the examples. The two operands are named the lvalue (on the left side) and rvalue (on the right side). An **rvalue** may be any C expression. An **lvalue** must be an expression that references a specific storage location, such as a variable name. Other **lvalues** will be covered in Chapter 8 (Arrays), Chapter 9 (Structures), and Chapter 10 (Memory and Pointers). A constant would not be a valid **lvalue**.

In addition to the = there are a number of compound assignment operators. For example:

```
a += 5;
```

The operator is **+=** and the operation is the same as:

```
a=a+5;
```

Simply put, these operators save you the time of typing the variable again if the same variable appears on both sides of the =. See assignment operators in Table 5.7.

Increment/Decrement Operators

C has dedicated operators to perform a common increment or decrement on data. For example, instead of:

```
x=x+1;
```

C has the **++** operator that does the same thing, like this:

Table 5.7 Assignment operator list.

=	Assignment
+=	Addition assignment
-=	Subtraction assignment
*=	Multiplication assignment
/=	Division assignment
%=	Modulus assignment
<<=	Left shift assignment
>>=	Right shift assignment
&=	Bitwise AND assignment
^=	Bitwise exclusive OR assignment
|=	Bitwise inclusive OR assignment

```
x++;
```

`++` is a unary operator and it can appear on either side of an expression, for example:

```
++x;
```

In this example the effect is exactly the same if the `++` is before or after the expression. However, like all expressions the `++` operator has a resulting value. The result of a `++` operation is different depending on where the `++` appears. If before the lvalue (prefix) then the result is the final value (after increment), otherwise (postfix) the result is the lvalue before the increment. We need an example to understand this:

```
x = 5;
y = ++x;
z = x++;
```

After all this we have 7 in `x` (twice incremented). There is a 6 in `y` because `x` was incremented before the result was returned. `z` will also have a 6 because the increment was after the result was returned. There is also a `--` that works in the same way to decrement. Increments and decrements are by 1 except in some special cases that will be covered in Chapter 10.

Other Operators

C has one ternary operator that can be a challenge to grasp. Here is the syntax:

operand1 ? operand2 : operand3

That `?` is a real question mark in the expression. Here is how it works:

If operand1 is nonzero the result of the expression is operand2
If operand1 is zero the result of the expression is operand3.

Here is an example:

```
average=count>0 ? sum/count : 9999;
```

In this case, average will be 9999 if the count is 0 otherwise the divide is done. Here is another example:

```
topay=total-total*(total>100 ? 0.15 : 0.10);
```

In this case we calculate a discount of 15% if the total is over 100 otherwise 10%.

Another unusual operator is the binary operator comma. For example:

```
x=y, z;
```

The comma operator evaluates both operands and always returns the second one. In this case **x** will be assigned **z**. The only reason to do this is if there is a side effect from the first operand being evaluated. For example:

```
x=y++,z;
```

In this case **y** would be incremented and **z** would be assigned to **x**. Remember an expression could be a function call and that could end up doing just about anything. You might only use this for the first or last expression in a for statement. Consider this:

```
for( a=0, b=0, c=0; a<=10; a++, b=b+10, c=c+3)
```

In this case you have three variables initialized and changing in the for loop. This was made possible by the comma operator.

There is another operator we have already used but did not identify as an operator. In C the keyword **sizeof** is considered a unary operator. It looks like a function call or macro but is in fact an operator. The parens used after **sizeof** are only required if the item after **sizeof** is a type name. The following are legal:

```
x = sizeof y;
x = sizeof(y);
x = sizeof(int16);
x = sizeof a+b;
```

Programmers almost always use the parens even where they are not required with **sizeof**. The **sizeof** operator simply returns the number of bytes in the type of the operand.

There are some additional operators we will not cover in detail until subsequent chapters (see Table 5.8).

Table 5.8 Additional operator list.

&	Address-of unary operator
*	Dereferencing or indirection unary operator
[ ]	Array indexing binary operator
.	Dot operator or structure member binary operator
->	Structure pointer binary operator

Sequence Points

C expressions are very powerful and by now it is clear that it is easy to create expressions that are difficult to understand.

Note that in an expression such as (A*B) + (C*D), the order the multiplies are done in is not specified by the standard but is up to the compiler. This makes a difference if one of the expressions has a side effect such as an assignment or function call. For example, consider this expression:

```
A=(B=C+2)+(C=B+1)
```

The result will be different depending on what order the two expressions inside parentheses are evaluated in.

Technically this is not valid code according to the C standard. However, the standard allows the compiler implementation to decide how to handle it and an error is almost never thrown. To understand why this is not valid you must understand the concept of a sequence point. The standard defines specific criteria for where in C code there are specific sequence points. One rule is a modified variable cannot be used between sequence points. The above code violates that rule. Rather than spending a lot of space explaining the specific rules for sequence points and how to identify them in the code it is better to simply say don't do crazy things like this. Even if you know how the compiler would handle the situation, the people that need to review and modify your code may not know.

Expression Examples

Assume you get a digital temperature from a sensor as an integer where the temperature is in Fahrenheit. You want to deal with the number in code as Celsius. Here is a simple conversion:

```
signed int16 tempF;
float tempC;

tempF=get_temp();
tempC=(tempF-32)*5/9;
```

When possible it is good to avoid float numbers. Here we have the formula where the result is in tenths of Celsius:

```
signed int16 tempF;
signed int16 tempC;

tempF=get_temp();
tempC=(tempF-32)*5*10/9;
```

Notice we multiply by 10 (to get tenths) before the divide. Mathematically it may seem the same but it is important here to do the multiplies first. This type of problem comes up frequently so be sure you understand why.

The above is also a good candidate for using fixed point.

Another issue that sometimes comes up is that the IEEE organization has a floating-point format that is very popular but not the same as the PIC®. The sign bit is in bit 31 instead of bit 23. The following formula will move the sign bit assuming the numbers are saved as integers for manipulation:

```
int32 ieee, pic;
...
ieee = (pic & 0x7FFFFF) | ((pic & 0xFF000000)>>1) | ((pic & 800000)<<8);
```

Summary

- C has mathematical, relational, bitwise, assignment, increment, decrement, and some special operators.
- C has unary operators that operate on one operand, binary operators that operate on two operands, and a ternary operator that operates on three operands.
- Like data, expression results have a type and that type can be converted to another type with a type cast.
- Some type conversions are automatic and do not require a type cast. They use the usual arithmetic conversion.

Exercise 5-1

Objective: Write a program using simple declarations, assignments, and expressions with **`printf`** to determine if the compiler does logical shifts or arithmetic shifts.
Requires: E3 module, USB cable, PC.

Steps/Technical Procedure	Notes
1. Write a program using simple declarations, assignments, and expressions with **`printf`** to figure out if the CCS C compiler does logical shifts or arithmetic shifts.	
2. Like the binary operators there is one math operator that does not work on a float variable. Figure out which one it is and test it with the compiler.	
3. All expressions have a result, some like the assignment operator have a side effect (the assignment). Write a program to find out what the result of an assignment expression is.	
4. In addition to **`printf`** there is another way to input and output data thanks to a C++ extension called streams. To output the variable x followed by a return/line feed you do this: cout << x << "\r\n"; For inputting a number into x you do this: cout << "Enter number: "; cin >> x; Using cout and cin write a program that enters from the user two int16 numbers and then outputs the sum of the two numbers.	
5. Write a program to output a requested bit in a byte. The user should enter the byte value and then your program will prompt for a bit number. It should then output the value of that bit, 0 or 1. The and operator can be used to isolate bits in a byte The shift operator can be used to shift a 1 a number of bits specified by a variable.	

Quiz

(1) Using the associate and precedence rules, in what order are the following operators evaluated?

```
a==b || c+d  || e*f
```

(a) == || + || *
(b) == + * || ||
(c) == * + || ||
(d) || || == * +
(e) This is an error, you cannot mix math and relational operators

(2) The following code is intended to convert three digits in ASCII to a integer. What is wrong with this code?

```
int16  value;
char  c1,c2,c3;
c1='3';
c2='2';
c3='1';
value = (c1-'0')*100+(c2-'0')*10+(c3-'0');
```

(a) You cannot use the subtract operator on characters
(b) Missing parens around the multiplication operations
(c) It would work only without each -'0'
(d) A type cast is required
(e) There is nothing wrong

(3) For all int8 variables, which of the following expressions will yield a different result?
(a) **x % 8**
(b) **x & 7**
(c) **(x<<5)>>5**
(d) **x>7 ? x-8 : x**
(e) All yield different results

(4) How could the following expression be described?

```
a==b==c==d
```

(a) True if all four variables are the same
(b) True if a is equal to b, and c is equal to d OR both are not equal
(c) True if c is equal to d and a is 1 and b is 1
(d) True if a is equal to b and c is 1 and d is 1
(e) This expression is illegal because the == is a binary operator

(5) How could the following expression be described?

```
(x & 1)!=0
```

(a) This does not make any sense although it is legal C
(b) True if x is not zero
(c) True if x is an odd number
(d) True if x is zero
(e) This is a syntax error

(6) According the C rules for expression evaluation, what is the value of x in the following expression?

```
x=(x=5,x+2)
```

(a) 0
(b) 1
(c) 5
(d) 7
(e) This is an illegal expression

(7) If x starts out as 0, what is the value of x in the following C expression?

```
x = ++++++--x;
```

(a) 2
(b) 258

(c) 3
(d) 0
(e) This is not a valid C expression

(8) Given 8-bit and/or 16-bit variables, an automatic C type conversion is not performed for which of the following expressions?
(a) **v16=v8**
(b) **v16>v8**
(c) **v8>v16**
(d) **v8*v8**
(e) **v16=v8*10**

(9) How many operators are in the following statement?

```
x += +10 * y;
```

(a) 1
(b) 2
(c) 3
(d) 4
(e) This is not a valid C statement

(10) What is x after the following expression?

```
x = sizeof( (int8)x * (int8)y )
```

(a) 16
(b) 8
(c) 1
(d) 2
(e) This is not a legal C expression

CHAPTER 6

Statements

Statements have been used in the past five chapters. A formal definition of all the statements in C is given in Table 6.1.

Key words used for statement identification cannot be used as normal C variable identifiers. They can be used for preprocessor identifiers; however, this is almost never right. The C standard name for conditional statements is "selection statements." Conditional statements are more widely used across other languages.

In addition, C allows statements to be labeled. The label appears before a **:** that is before the statement. For example:

```
a=1;
lab1: b=2;
c=3;
```

The lab1 is a label for the b=2 statement. The label can be used in the goto statement to force the program to jump to a specific statement. The rules for labels are the same as for any other identifier. Labels are always local to a function in scope.

There are two special label formats that are only used inside a switch compound statement. The case and default label syntax will be covered with the switch statement later in this chapter.

The important thing to remember in statement syntax is where the statement begins and ends. For example, we have used the for statement before and know that the loop is only a single statement. Usually there is more than one statement in a loop, so a compound statement **{ }** must be used. The same consideration needs to be applied to all the statements in Table 6.1 where a *stmt* is indicated as part of the statement. The *stmt* must be replaced with a valid C statement.

Pay close attention to where the **;** is in each statement, and when other statements are a part of a statement, remember that each statement must follow its own rules.

Embedded C Programming. http://dx.doi.org/10.1016/B978-0-12-801314-4.00006-5

Table 6.1 Statement list.

Statement	Description
;	Null statement
expr;	Expression statement
if (cond) stmt;	Basic conditional
if (cond) stmt; else stmt;	Two-path conditional
switch(expr) stmt;	Multi-path conditional
for(expr; expr; expr) stmt;	Complex loop
while (cond) stmt;	Simple loop with the condition checked first
do stmt while (cond);	Simple loop with the condition checked last
break;	Jump out of loop
continue;	Jump to top of loop
goto label;	Jump to label
return expr;	Return from function, the expression is optional
{ stmt list }	Compound statement
declaration;	A data declaration

Statement Definitions

if Statement

```
if (condition)
    statement;              //single statement
   OR
if (condition) {
    statement1;
    statement2;             //block of statements
    . . . .
    statementn;
}
```

If the condition is true (nonzero), the statement(s) is executed.
If the condition is false, the statement(s) is not executed.
In either case, execution continues with the line of code following the statement(s).
Note that if(condition) and statement(s) are all part of the if statement (see Figure 6.1).

Also remember that true really means the expression is nonzero.

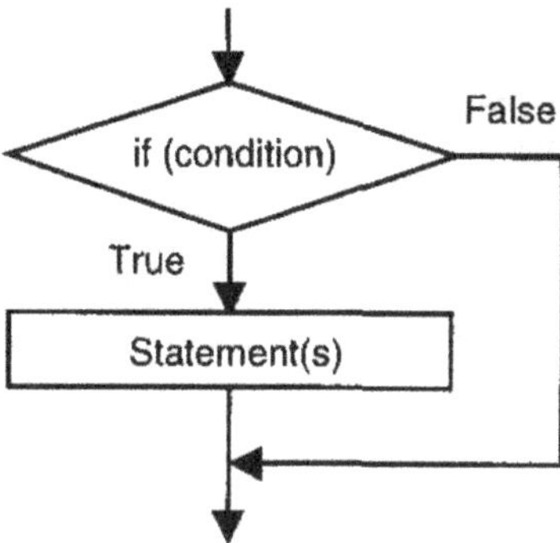

Figure 6.1: if statement flow chart.

The relational operators (==, >, <, >=, <=, !=) are frequently used as the basis of conditions in *if* and *if/else* conditional statements to direct program flow. The logical operators && (and), || (or), and ! (not) may also be used as the basis of conditions in *if* and *if/else* conditional statements to direct program flow.

Some examples:

```
if( x > 99)                    // Don't allow x to be greater than 100
  x=100;
if( error ) {                  // If error is not zero then report it and clear the count
  printf("An error has occured\r\n");
  n=0;
}
if( (c='A') || (c='E') || (c='I') || (C='O') || (C='Y') )
  printf("Vowel");
if( (c>='0') && (c<='9') )
  printf("Number");
if( (c=toupper(c))=='W' )
 printf("W command");          // This example uses both = and == in the condition. This is
                               // an expresion with a side effect. The toupper() function converts
                               // a character to upper case. In this case we modify c itself to
                               // be upper case and then check to see if it is a W.
                               // The condition is true if c comes in as 'w' or 'W' and it has
                               // the side effect of converting c to uppercase.
```

An if conditional statement may, as an option, contain an else clause (see Figure 6.2).

```
if (condition)
  statement(s)1;

else
  statement(s)2;
```

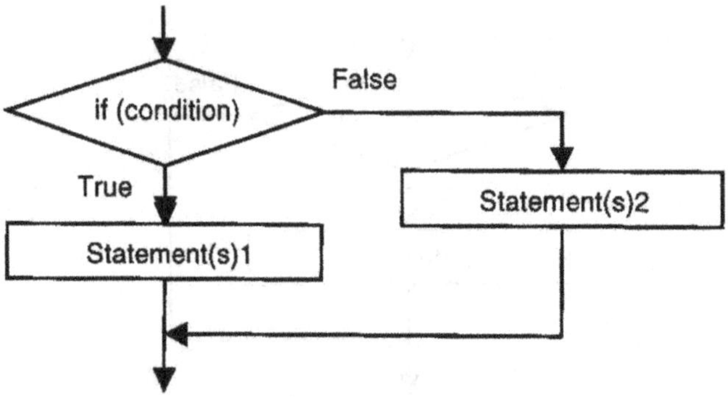

Figure 6.2: if/else statement flow chart.

If the condition is true, statement(s)1 is executed and statement(s)2 is not executed. If the condition is false, statement(s)1 is not executed and statement(s)2 is executed. In either case, execution continues with the line of code following statement(s)2.

if/else statements may be cascaded to create a hierarchy for decision making (see Figure 6.3).

```
if (condition1)
  statement(s)1;

else if (condition2)
  statement(s)2;

else if (condition3)
  statement(s)3;

else
  statement(s)4;
```

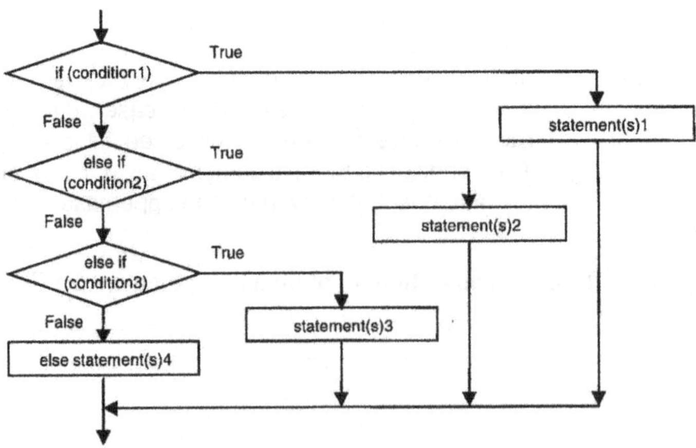

Figure 6.3: Nested if/else statement flow chart.

For this sequence of if/else statements, only one will be executed and all that follow will be skipped over. The first if/else statement in the sequence that evaluates as TRUE will be executed. If conditions 1, 2, and 3 all evaluate as FALSE, statement 4 will be executed.

Some examples:

```
if( error ) {                // If error is not zero then report it and clear the count
  printf("An error has occured\r\n");
  n=0;
} else
  printf("No errors\r\n");

if( x > 99)                  // Force X to be 20 to 100
  x=100;
else if (x < 20 )
  x=20;
```

while Loops

```
while (condition)
  statement(s);
```

Repeat code until condition becomes false (see Figure 6.4).

1. Condition is evaluated first.
2. If condition is true, statement(s) is executed and execution loops back.
3. If condition is false, the while statement is terminated and execution passes to the code following the while loop (if there is any).
4. The code following the while loop (if there is any) may not be executed at all if condition remains true.

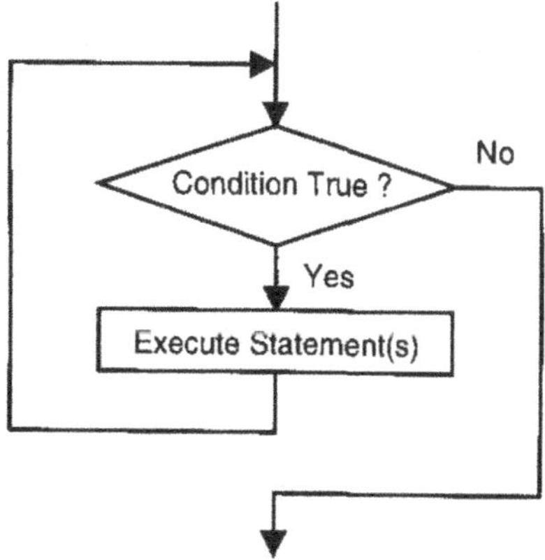

Figure 6.4: while loop flow chart.

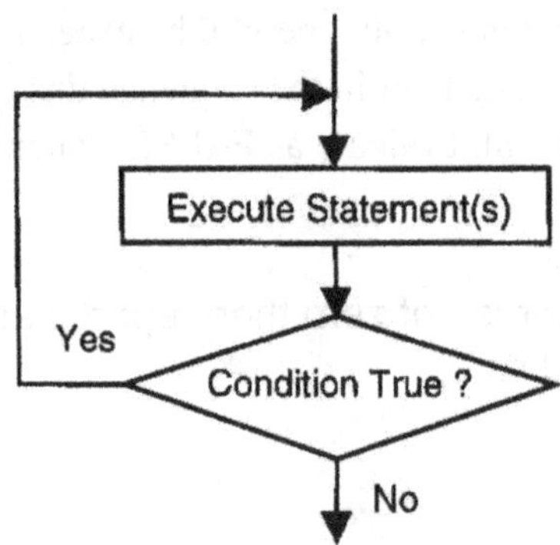

Figure 6.5: do while loop flow chart.

Use the while loop for simple situations where a simple condition such as a switch being open or closed controls how long to stay in the loop. The previous form is used when the statements might never be executed (if the condition is initially false). The following form is used when the statements are always executed at least once (see Figure 6.5).

```
do
  statement(s);
while (condition);
```

1. The statement(s) is executed first.
2. Condition is evaluated.
3. If condition is true, execution loops back and the statement(s) is executed again.
4. If condition is false, while statement is terminated and execution passes to the first statement following while (condition).
5. The statement(s) in braces will be executed at least once.

A special form of the while loop is the forever loop. This is a loop that never ends. It is common to see this in an embedded application in the main program. Unlike a PC program, an embedded program may just run forever (or as long as it is powered up). It looks like this:

```
while(TRUE) {
  gather_data();
  process_data();
  output_data();
}
```

The following form simply stops execution in its tracks (1 and true are interchangeable):

```
while(1);    // This is microprocessor suicide, it never exits this line
```

Some examples:

```
while( sum<10000  && !error ) {        // Loop until a sum of 10000 or
                                         // error is false
   value = get_data();
   if( value!=0 )
      sum +=value;
   else
      error=TRUE;
}

do {
   cout << "Enter number: ";
   cin >> x;
   cout << "Square is " << x*x << "\r\n";
} while (x!=0);                        // Loop terminates when user enters 0

count=1;                               // Outputs the numbers 1 through 10
while( count <=10 ) {
   cout << count << endl;
   count++;
}

count=0;                               // Outputs the numbers 1 through 10
while( ++count <=10 )
   cout << count << endl;

count=1;                               // Outputs the numbers 1 through 10
do
   cout << count << endl;
while( count++ <=10 );

count=0;                               // Counts the number of binary 1's
bit=-1;                                // in data
while( (data!=0) && (bit<(sizeof(data)*8)))
   if( data & (1 << ++bit) )
      count++;

root=0;                                // finds the square root on an integer
do { // if it is a whole number
   root++;
} while (root*root<data);
```

The CCS C compiler has some functions to read the state of an I/O pin and to set a pin high or low. These functions use constants defined in the device header file to identify each pin. The input(), output_high(), and output_low() functions are used in the following example (see Figure 6.6).

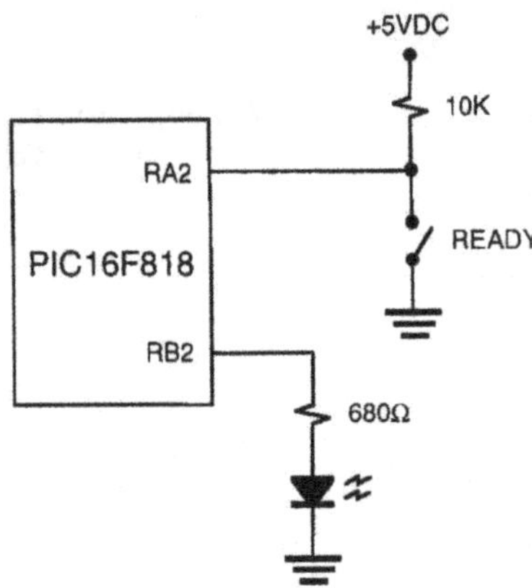

Figure 6.6: PIC® MCU schematic with button and LED.

On reset, the "ready" switch is open and pin A2 is pulled up to +5V via a pull up resistor, pin A2 = 1.

Following is a simple while loop to wait for the switch to be pressed. Normally the switch reads 1 (true or high) and when pressed it goes to 0 (false or low). To detect a press we wait while the switch is not pressed (high):

```
while (input(PIN_A2)==1);
```

This used the math operator == and we could have instead simply done the following:

```
while (input(PIN_A2));
```

A program to turn the LED on when the switch is pressed looks like this:

```
void main(void) {
   while (input(PIN_A2));          //go low? wait for ready switch to close
   output_high (PIN_B2);           //indicate ready switch closed
}
```

Here is a variation of the program so that each time the switch is pressed and released the LED will blink three times. This shows a loop nested inside another loop. It also uses another CCS C built-in function to delay for a certain amount of time.

```
void main(void) {
   while(TRUE) {
      while (input(PIN_A2));       //Wait for press
      while (!input(PIN_A2));      //Wait for release
      count=0;
      while(count<3) {
         output_high (PIN_B2);     // LED on
         delay_ms(1000);           // Wait a second
         output_high (PIN_B2);     // LED off
         delay_ms(1000);           // Wait a second
         count++;
      }
   }
}
```

for Loop

for (start expression; test expression; loop expression) statement(s);

```
for(i=0;i<=n;i++)
start expression:  i = 0
test expression:   i <= n
loop expression:   i ++
```

The for statement is a very flexible version of the traditional for statement in other high-order languages. Typically the primary purpose of a for statement is to execute a loop a specific number of times. The C **for** can do that and much more. It can manage any number of counters and have multiple conditions to exit the loop (see Figure 6.7).

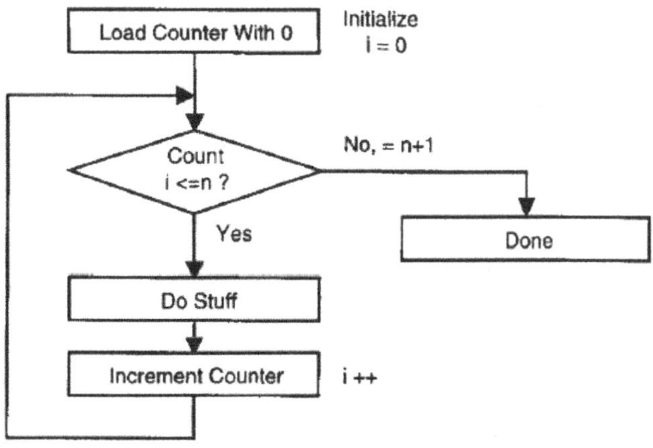

Figure 6.7: for loop flow chart.

The start expression is evaluated only once before the loop begins. The result of the start expression is ignored.

The test expression is tested at the top of every loop and when nonzero the loop is executed, otherwise the statement terminates.

The loop expression is evaluated at the bottom of the loop and the result is ignored.

The same loop can be coded with the while loop as follows:

```
start expresion;
while( test expresion ) {
  statement(s);
  loop expresion;
}
```

It is considered bad form to modify a variable that appears in the loop expression inside the loop statements.

Some examples:

```
for( i=1; i<=10; i++ )                // prints 1 to 10
  cout << i << endl;

for( i=0; i<=100; i+=7 )              // prints all multiples of 7 less than 100
  cout << i << endl;

for( ; ; ; ) {                        // loops forever, another form of a forever loop
  input_data();
  process_data();
}
```

Jump Statements

These statements cause the normal flow of execution to be diverted to another point. The break and continue could be implemented with if statements in all loops; however, breaks and continues will make the code look cleaner and more readable. C allows both to be used outside loops and in that case they do nothing. In both cases the statement jumps only with respect to the innermost loop. If you have a loop inside a loop then only one loop is affected; for example, on a break only the inner loop terminates.

break statement—causes control to jump out of a loop or switch statement (see Figure 6.8).

continue statement—causes control go to the beginning of a loop statement (see Figure 6.9).

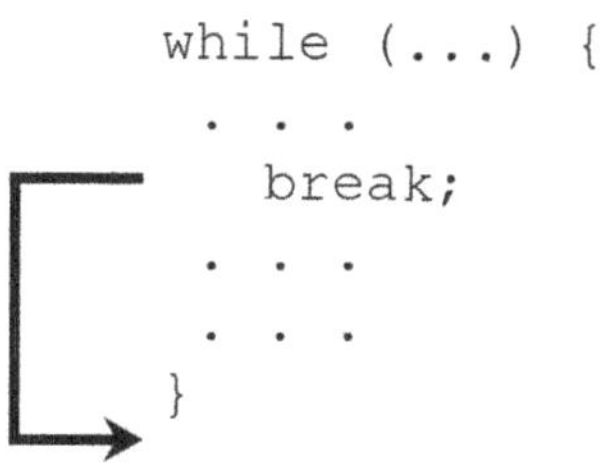

```
while (...) {
  . . .
    break;
  . . .
  . . .
}
```

Figure 6.8: break statement illustration.

```
while (...) {
  . . .
    continue;
  . . .
  . . .
}
```

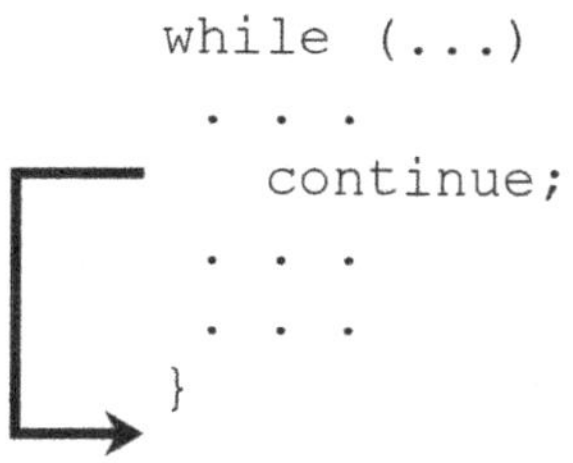

Figure 6.9: continue statement illustration.

return statement—used to return from a function call (more in Chapter 7, Functions). The return statement is used to exit a function. It can also pass the value of the function out to the caller if the function is not void. If a function has no return directive, then a return is done after the last statement of the function is executed. A function can have any number of returns; however, some coding standards will say a function should only have one return and it should be at the end of the function. This may make the code easier to understand.

```
return;
```

or

```
return data;
```

goto—execution branches to location identified by label within the same function. C purists do not like the use of goto statements and go all out to avoid using them. It can make the code very difficult to understand.

```
goto label;
....
....
label: statement;
```

Simple example:

```
while(TRUE) {
    cout << "Enter number: ";
    cin >> x;
    if( x==0 )
        break;
    if( x>255 ) {
        cout << "Numbers must be one byte\r\n";
        continue;
    }
    cout << "Square is " << x*x << "\r\n";
}
```

switch/case Statement

```
switch (expression) {
  case 0: statement(s);
    break;
  case 1: statement(s);
    break;
  default: statement(s);
    break;
}
```

When a variable can be one of several values, switch/case may be used instead of several **if/else** statements. As an option, a default may be used that takes care of any cases not otherwise included. If a default is not used and there is no match between expression and the cases given, execution passes to the first statement following the switch statement's closing brace.

Expressions must evaluate to an integer or a char. The CCS C compiler also allows character strings to be used in a switch/case; however this is not allowed in standard C. In this example, expression should evaluate to 0 or 1. If not, the default statement(s) will be executed (see Figure 6.10).

Notice the case items are a special syntax of the label. The statements, unlike the loops, do not need the **{ }** to group them because they are a long list of statements with labels to identify the branch points. The break is needed to prevent the next group of statements from being executed.

The case items need not be in order nor consecutive. They do, however, need to be unique.

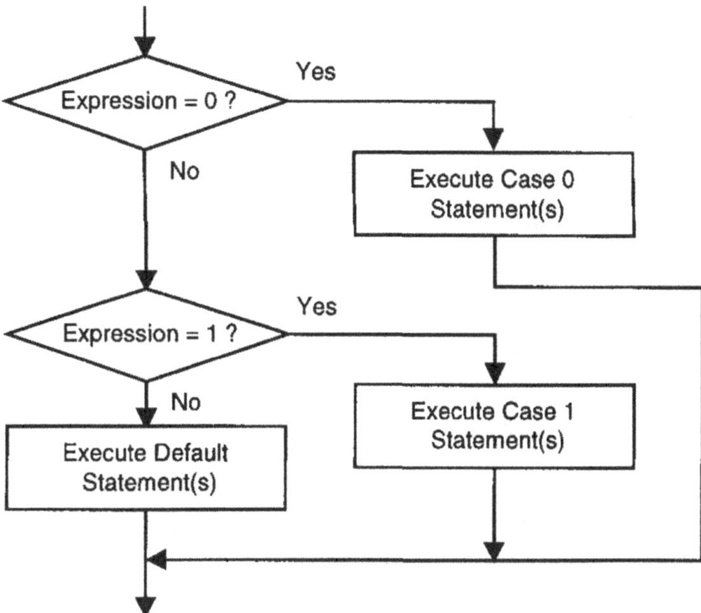

Figure 6.10: switch statement flow chart.

An example:

```
switch( cmd ) {
   case 'W':
   case 'w':     perform_write() ;
                 break;
   case 'R':
   case 'r':     perform_read() ;
                 break;
   default:      printf("Unknown Command\r\n");
                 break;
}
```

Side Effects

In the if examples above we showed an example of an expression with a side effect. Some programmers can sneak these in almost anywhere. Use of the assignment operators in conditional expressions or using more than one assignment operator in an expression statement can be confusing. When reviewing code be sure to watch out for this. When writing code it may be best to avoid those constructs unless it is very clear what is going on. Here are some more examples:

a=b=c=d=e=f=0;
if(c1=getc() == c2=getc())
point=basevalue *(offset+=gapsize);

The ++ and -- always have a side effect and these are commonly used inside expressions. Again when reviewing code be aware the lvalue associated with those operators will change.

Nesting, Indentation, and Use of Braces

Statements inside statements are clearly allowed. There is no limit to the amount of nesting that can be done. Be careful to match up where breaks and continues will go.

Nesting statements using conditionals and loops can be complex to read. Indentation can be used to help identify for the human reader where the loops and conditional branches are. Consider what the following code would look like without indentation. Usually there is an indentation change when the statement is part of a new control branch or loop. A specific company coding standard may specify more specifics on the indentation. The C compiler does not care.

Likewise, the use of { and } for a single statement, and if the { is at the end of a line or under the first character of the line, is a style preference. Almost always you will want to line up the } with the first character of the loop or conditional start.

```
if (c >= '0' && c <= '9') {
   if (c == '0' && (s[index] == 'x' || s[index] == 'X')) {
      base = 16;
      index++;
   }
   c = s[index++];
}
if (base == 10) {
   while (c >= '0' && c <= '9') {
      result = (result << 1) + (result << 3);  // result *= 10;
      result += (c - '0');
      c = s[index++];
   }
} else if (base == 16) {
      c = toupper(c);
      while ((c >= '0' && c <= '9') || (c >= 'A' && c <='F')) {
         if (c >= '0' && c <= '9')
            result = (result << 4) + (c - '0');
         else
            result = (result << 4) + (c - 'A' + 10);
         c = s[index++];c = toupper(c);
      }
}
```

Be careful with indentation. Notice the following code does not work as the indentation implies:

```
if (base == 10) {
   while (c >= '0' && c <= '9')
      result = (result << 1) + (result << 3);  // result *= 10;
      result += (c - '0');
      c = s[index++];
}
```

The CCS IDE can automatically reform your source code according to some user-defined rules. Highlight your source, then use EDIT>FORMAT SOURCE. The same IDE ribbon allows you to easily indent or unindent selected text. These features will help you to keep your code tidy.

Design Documentation

Before coding larger programs and to document for others the program algorithms, some form of documentation should be generated. One popular method to describe the algorithms in a function is to use flow charts. This chapter provides the common flow chart elements for each statement that controls program flow. Typically these elements would be combined to fully describe a single function. Flow charts tend to be easy to read for programmers and non-programmers alike. They do however take a lot of time to generate and update when changes are needed. The CCS C IDE has a built-in flow chart editor to help in this process. An example flow chart is shown in Figure 6.11.

An alternative used by programmers is pseudo-code. Pseudo-code is a cross between code and text. The purpose is to describe an algorithm to the detail required by a human to understand but not to the detail for a compiler to use. There are no specific syntax rules and indentation is used to replace items like **{** and **}** . The following is example pseudo-code.

```
Check_speed function:
   If the state is running
      Get the current speed
      If the speed is over the target OR the heat is over MAX_HEAT
         If we are at the minimum speed
            Light the fault LED
            Set state to FAULT
            Shut motor off
         Else
            Reduce speed
         Else
            If speed is less than target
               Increase speed
```

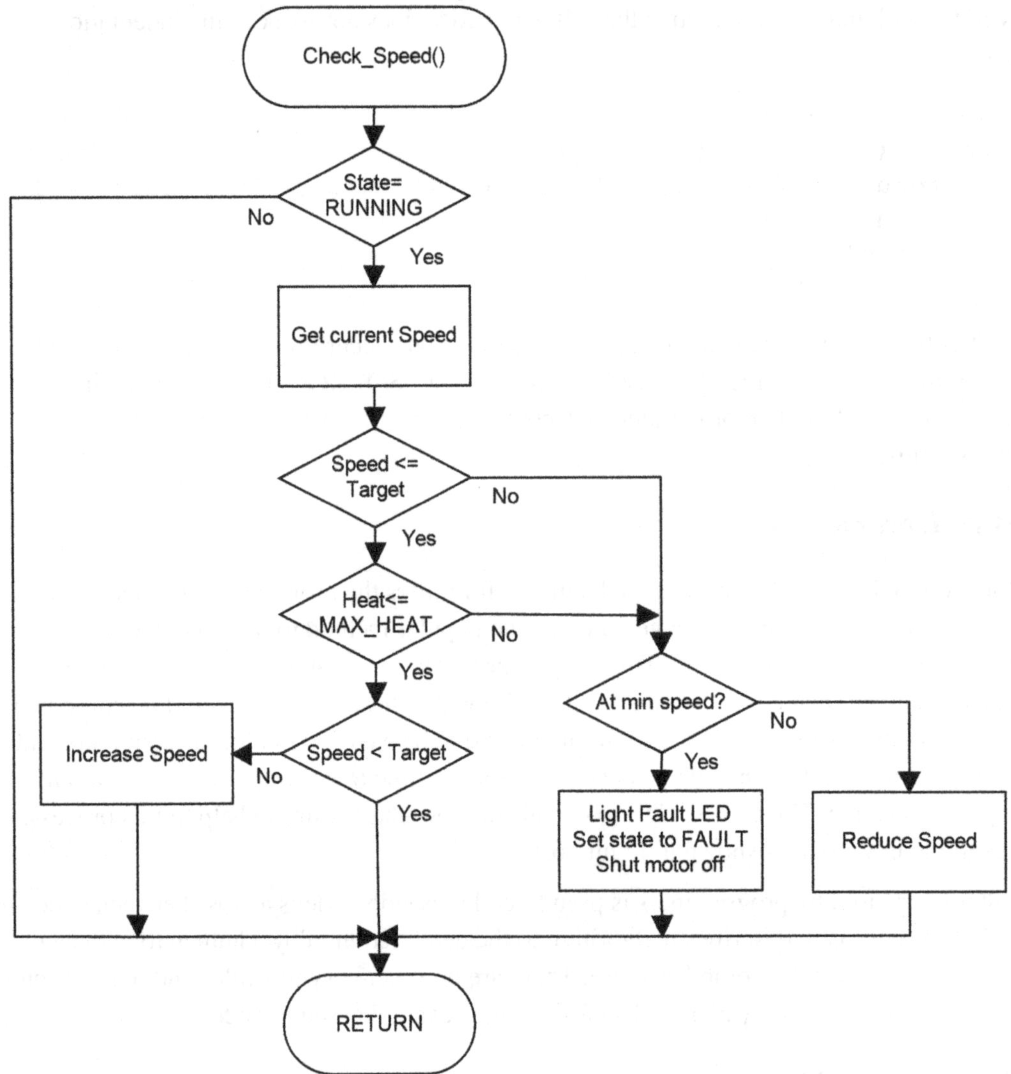

Figure 6.11: Example function flow chart.

Program Complexity

Code metrics are often used to assist software project management. Established metrics help to determine how complex a program is so one can know what to expect for development time and ease of maintenance.

One type of metrics used is the Halstead metrics, based on the unique and total number of operators and operands in the program. Halstead based his metrics in psychology and how

Higher volume, lower difficulty	Lower volume, higher difficulty
A+=X B+=X C+=X D+=X E+=X F+=X G+=X H+=X	A=B*C+B*A-B/D-B/C
Volume = 156 Difficulty = 1.9 Effort = 296	Volume = 89 Difficulty = 3.6 Effort = 320

Figure 6.12: Halstead metrics example.

the human brain is able to keep track of what makes up the total program. Consider the examples in Figure 6.12.

In general, the difficulty is greater as there are more unique operators and operands. Function calls are considered as an operator; however, Halstead does not take into consideration control statement nesting.

To properly account for complexity of control statement nesting, the cyclomatic complexity metric is frequently used. In short, it measures the number of possible paths through the code. A simple if statement has two paths. When adding an if to a then else, the result is four paths. This directly correlates to the number of test cases needed to fully test a program.

The CCS C compiler IDE measures and calculates the most common metrics used to measure code, including Halstead's complexity metrics, cyclomatic complexity, and the maintainability index. The statistics file will have in it the total time estimated to implement, estimated number of initial bugs, and how hard the code will be to maintain.

These metrics may be used to identify functions or programs that are in need of simplification. They may also be used to evaluate productivity and can help in future project estimating. This is mentioned at this point in the book so the designer can be aware of what constructs are considered more complex and harder to implement and maintain.

Summary

- C has 14 standard statement types that are executable instructions within functions.
- Some statements have additional statements nested within.
- Conditional, loop, and jump statements can control the flow of program control dictate what statements are executed.
- The indentation of statements and the level of nesting can affect the readability and maintainability of the program.
- Various standard metrics are available to determine overall code complexity.
- Key words used for statement identification cannot be used as normal variable identifiers.
- Labels may be inserted anywhere in the code a statement can appear, to be used as a marker for other statements.
- Flow charts and pseudo-code are methods to document the control flow algorithms before and/or after coding.

Exercise 6-1

Objective: Become familiar with the use of the basic C statements. In addition, understand how to generate flow chart documentation.
Requires: E3 module, USB cable, PC.

Steps/Technical Procedure	Notes
1. Write a program to enter two numbers from the user and then print out all odd numbers inbetween and including the entered numbers. The numbers can be entered low or high first and output in numeric order. When done the program should prompt for another set of numbers. Use the flow chart editor to generate a flow chart to document the control flow for this program.	
2. Write a program that prints all the room numbers for an apartment building. Print each floor on one line. The building has 14 floors numbered 1-15 skipping 13. Each floor has 10 rooms A-J except the first floor that has only E-J.	
3. Write a program that prompts the user to enter a number 0-9 and then the program prints out the text string for the number. For example, an entry of 5 should output "Five." When done the program should prompt for another number.	
4. Modify the above program to allow numeric entry of up to 9999. A value of 1234 should output "One Thousand, Two Hundred and Thirty Four." A value of 5 should still output "Five." Before coding this new program, use pseudo-code to show the algorithm you will be using.	

Quiz

(1) How many different paths are there through the following code?

```
if( aa )
   if( bb )
      cc=1;
      else
         if( dd )
            cc=2;
         else
            cc=3;
```

(a) None
(b) 2
(c) 3
(d) 4
(e) This is not legal C

(2) Why is the following code illegal?

```
for( int n=0, float x=1; x<=10; x+=1,
   for(int y=1; y<=10; y++) n++)
      n++;
```

(a) Floating-point numbers cannot be used in a for statement
(b) Statements are not expressions
(c) The code is too hard to read and maintain
(d) The += cannot be used on a float
(e) It's not illegal, n will be 100 upon exit

(3) For the following, what are the conditions where x will equal 3?

```
x=0
if( e1 )
   if( e2 )
      x++;
   else
      if( e3 )
         x+=2;
      else
         x=3;
```

(a) e1 false, any value for e2, e3
(b) e1 true, e2 false, e3 false

(c) el, e2, and e3 all true
(d) el true, e2 true, e3 false
(e) It is not possible for x to be 3

(4) What is the value of x after the following code executes?

```
x=0;
for(i=1; i<=5; i++)
    for(j=1; j<=5; j++,x++)
          if( j==i )
             break;
```

(a) 0
(b) 5
(c) 24
(d) 25
(e) The code is not legal C

(5) What is the value of x after the following code executes?

```
x=0;
y=2;
switch( y ) {
    case 0 : x+=1;
    case 1 : x+=2;
    case 2 : x+=4;
    case 3 : x+=8;
}
```

(a) 0
(b) 1
(c) 4
(d) 12
(e) 15

(6) Of the continue, break, return, and goto statements, which two might appear on two consecutive lines and make sense?
(a) continue followed by a break
(b) break followed by a continue
(c) any of them followed by a return
(d) continue followed by a goto
(e) No combination makes sense

(7) What is the problem with the following code?

```
y=0;
do
   while( y< 3 )
      y++;
while( y<5 );
```

(a) Not legal C, it is missing `{ }`
(b) Not legal C, whiles cannot be nested inside do-whiles
(c) Infinite loop, the first do matches with the first while and nothing happens inbetween.
(d) Infinite loop, once y reaches 3 it never increments to 5 to exit the outer loop
(e) No problems

(8) What is the difference between the following two code segments?

(a) The B code has more statements
(b) The B code is easier for a compiler to optimize
(c) The A code will work with floats and the B code will not

```
A.if( x==1 )  fone();
  else if(x==2)  ftwo();
  else if(x==3)  fthree();
  else ferror();

B.switch( x ) {
     case 1: fone(); break;
     case 2: ftwo(); break;
     case 3: fthree(); break;
     default: ferror(); break;
  }
```

(d) All of the above
(e) There is no difference

(9) What is the correct way to stall a program until a signal goes from low to high and then back to low (pulse detect)?

(a)
```
while(input(SIGNAL));
while(!input(SIGNAL));
while(input(SIGNAL));
```

(b)
```
while(!input(SIGNAL));
while(input(SIGNAL));
```

(c)
```
while(!input(SIGNAL))
while(input(SIGNAL));
```

(d)
```
do {
    while(!input(SIGNAL));
} while(input(SIGNAL));
```

(e)
```
do while(input(SIGNAL));
do while(!input(SIGNAL));
```

(10) Is the following code legal C?

```
#define  WAIT_FOR_SIGNAL_HIGH  while(!input(SIGNAL));
#define  WAIT_FOR_SIGNAL_LOW   while(input(SIGNAL));
...
   for(i=1;i<=5;i++) {
      WAIT_FOR_SIGNAL_HIGH;
      WAIT_FOR_SIGNAL_LOW;
   }
```

(a) Not legal, you cannot use a preprocessor identifier in a #define definition
(b) Not legal, you can only have expressions, not statements, in a #define
(c) Not legal, there would be two ; in a row at the end of the two wait lines
(d) Not legal, you cannot have a loop in a preprocessor directive
(e) Yes, it is legal C

CHAPTER 7

Functions

A *function* is a routine that performs a task. A function may come with the C compiler you are using or you may write one yourself. The CCS C compiler comes with a lot of functions designed to do specific tasks such as writing to a port, creating a time delay, setting up an A/D converter, etc. The fact that these built-in functions are provided will save a lot of work and your projects will be completed sooner.

Data may be passed to functions as arguments (see Figure 7.1).

A function may or may not return data. C allows for a single return value from a function; however, when we learn about pointers we will see there are ways to get a function to pass more data out than the normal return value.

You may write a function to perform a specific task, test it, and store it away so it can be retrieved and used the next time you are writing code that needs that function. Putting related functions in a #include file is one way to easily reuse those functions in multiple programs.

main() FUNCTION

All C programs contain a *main function* that is the first function executed when the program starts. The name of this function is always main().

The parentheses that follow a function name indicate that it is a function (see Figure 7.2).

Data is sometimes passed to functions as arguments contained within the parenthesis.

In this example, there are none. Braces `{ }` indicate what is included in the function. In the case where a function has no parameters, and no return value, it is considered good form to define as follows (see Figure 7.3).

Function Definitions

C is designed to encourage modular or structured programming. Programs are constructed using modules called functions. Each function performs a specific task. In assembly language and some others, they are called *subroutines*.

Embedded C Programming. http://dx.doi.org/10.1016/B978-0-12-801314-4.00007-7

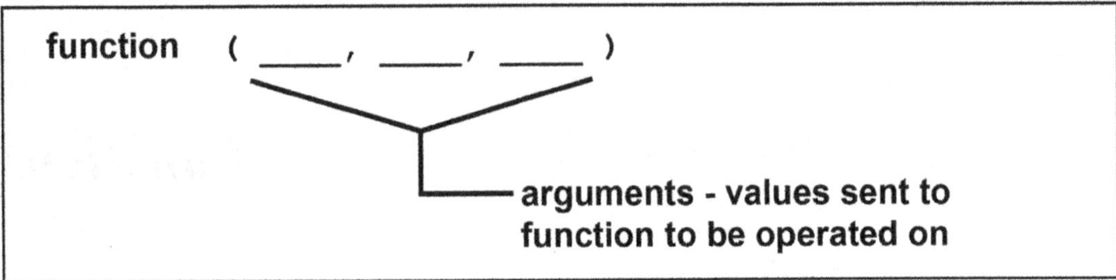

Figure 7.1: Function call illustration.

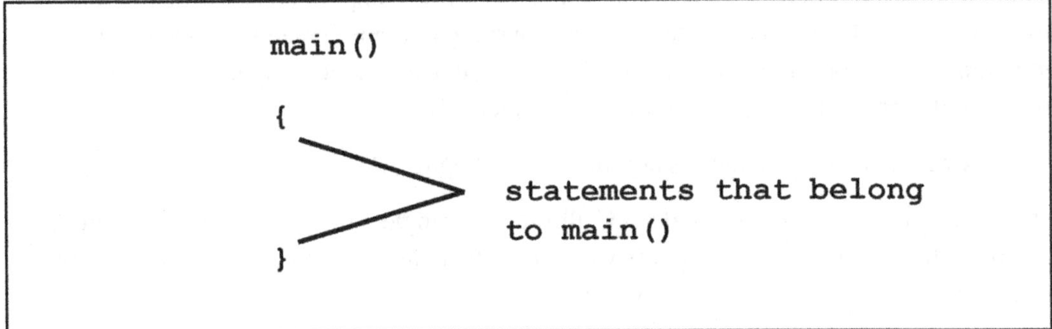

Figure 7.2: Function body illustration.

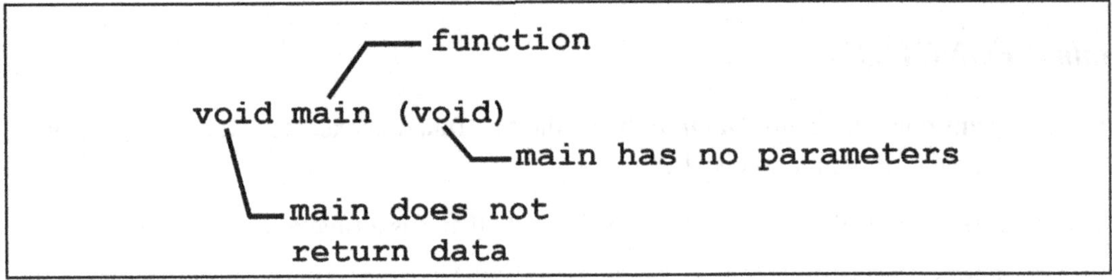

Figure 7.3: Function call that does not get or return specific data

The main function calls other functions to perform tasks. Further, any function can call any other function.

The functions that you create must be defined before they are used. That means you will typically see a layout like this:

```
funct1() {
    _ _ _ _
    return;                 //return to calling function, main
}


funct2() {
    _ _ _ _
    return;                 //return to calling function, main
}

funct3() {
    _ _ _ _
    return;                 //return to calling function, main
}


main()   {
    funct1();               //calls function 1
    funct2();               //calls function 2
    funct3();               //calls function 3
}
```

The return statement sends execution back to the calling function.

For other than very simple programs, the main function is used simply to call each of the functions designed to perform the tasks that the program is required to accomplish.

If reading the code bottom up bothers you, C has a way to identify a function before it is fully defined. This is called a *prototype*. The prototype simply identifies the function, the return value, and the parameters without listing the code in the function. Instead of the **{ }** after the function, there is a simple **;**. Here is an example:

```
funct1();
funct2();
funct3();

main()   {
    funct1();               //calls function 1
    funct2();               //calls function 2
    funct3();               //calls function 3
}


funct1()  {
```

```
   _ _ _ _
   return;              //return to calling function, main
}

funct2() {
   _ _ _ _
   return;              //return to calling function, main
}

funct3() {
   _ _ _ _
   return;              //return to calling function, main
}
```

The void for the return value and parameters are optional; however, some compilers will issue a warning if you omit them as we did above. Be aware if a function does not have void for a return value, C assumes the return value to be an *int*.

Function Parameters

The parameter list for a function looks a lot like a variable declaration. In fact, each variable identified in the parameter list becomes a local variable. Standard C passes data to function using a method called *pass by value*. This means when a function is called the arguments in the call are copied to the local variable parameters inside the function. The expressions in the function call are called *arguments*. For example:

```
void  funct1( int&a, long b ) {
   ...
}
void main(void) {
   int x;
   x=3;
   funct1(x,5);
}
```

In this case when **funct1()** is called there will be two local variables that may be accessed, called **a** and **b**. **a** will have a value of 3 and **b** will have 5 for this one function call. It may get other values on other calls. The normal arithmetic conversion of the types of the source to destination variable is done here. It works just like a normal assignment operator.

Since **a** and **b** are variables like any other, you may write to those variables inside the function. For example you could do **a=7;** inside **funct1()**. If this is done, however, when the function returns the variable, **x** does not change. A 3 was copied to **a** when the function was called but there is no copy back on the return.

In both the parameter list and argument list, each item is separated by a comma. If no type is indicated the type is assumed to be *int*. The following are examples of parameter lists in a function declaration:

```
funct( int a )
funct( signed long a, float b )
funct( a,b,c,d )
```

The last example where no types are shown is allowed (default int type); however it is not considered good form. Note that not all type qualifiers are allowed in a parameter list. For example, static is not permitted here.

Some example argument lists (the call):

```
funct( 1 );
funct( a, b );
funct( x+y-3 );
funct( a, (b,c) );      // Two arguments, second expression uses comma
                        // operator for no good reason
```

Advanced Features

Compatibility Notes

The following three topics, reference parameters, default parameters, and overloaded functions, are C++ language extensions. Most standard C compilers will not support them.

All three are part of the CCS C compiler.

Reference Parameters

C++ added another method to pass arguments, called *pass by reference*. When you pass by reference the data from the argument is not copied, but rather a link to the original variable is passed and the original variable is actually used in the function. In this case, the function can modify the actual variable that appears in the calling argument. The `&` symbol is used between the type and identifier to indicate a reference parameter. The following is an example:

```
void  funct1( int & a, long b ) {
   a=a+b;
}

void main(void) {
   int x;
   x=3;
   funct1(x,5);
}
```

In this case **x** will equal 8 after the function is called. Note that if we made **b** a reference parameter in this example, there would be a compile error because you must pass an *lvalue* (not constant) for a reference parameter.

Default Parameters

C++ added another language extension that allows arguments to be omitted by the caller. When an argument is omitted there is a default value that is used. For example:

```
void  funct1( int & a, long b = 1 ){
   a=a+b;
}

void main(void) {
   int x;
   x=3;
   funct1(x);
   funct1(x,2);
}
```

In this case **x** is 4 after the first call and 6 after the second. Note that is not legal to call a function without an argument, unless the argument omitted has a default value (using the = symbol). Default parameters must appear after all mandatory parameters in the list.

Overloaded Functions

One more feature from C++ that is helpful is called overloaded functions. This is a powerful feature that can be easily misused if one is not careful. This feature allows the programmer to define two functions with the same name but with different parameters. The compiler decides what function to call based on the argument list. For example, assume we want a function that adds two numbers but does not allow a result over 100. Furthermore, we want to do this for integers and floats.

Using overloading we could do this:

```
int  add100( int a, int b ) {
   if( (a+b)>100)
      return 100;
   else
      return a+b;
}

float  add100( float a, float b ) {
   if( (a+b)>100)
      return 100;
   else
      return a+b;
}
```

If you call **add100()**, with floating-point numbers the second function is called, and with integers the first is called. If you call it with one float and one integer then the integer is converted to a float and the second function is called. It should be noted that overloaded functions need not have the same number of parameters. In fact a difference in the parameter count can be used to select the right function instead of the types.

Some older-style programmers will only specify the type in a prototype and leave the type off in the full declaration. This does not work well for compilers that support overloading. If you are porting older code that uses this style in the code, then either add the types or disable the compiler overloading feature (there is a pragma for that).

Return Values

Inside the function when a return is encountered with an expression, the expression is converted to the type of the function and the value is copied to the expression of the caller. Execution of the function terminates at this point. Functions can have multiple returns; however, as soon as one executes the function is done. Here is an example:

```
int  min( int a, int b ) {
   if( a< b )
      return a;
   else
      return b;
}
...
points = min( entries, 100 );
```

In this case, entries are copied to **a** and 100 is copied to **b**. In the function if **a** is less than **b** then **a** is returned and assigned to points, otherwise **b** is assigned to points. In effect points will get entries unless entries is over 100 and in that case it gets 100.

If a function does not have a return with an expression then the result returned to the caller is undefined (garbage). Many compilers will give you a warning about this. If a void function is used inside an expression you should get an error.

A caller does not need to use the function result. For example, this is legal but in this case useless:

```
min(entries, 100);
```

It is common for functions to do something useful and they return some kind of status result. The result may not be needed in all cases where the function is called, so it is perfectly normal to make the call without looking at the result.

Inline Functions

Assume you have a simple function that is called a lot. In fact the function may be so simple that you think just the act of calling and returning from the function may involve more work than what is in the function. Almost all C compilers allow a function to be tagged as inline.

This means the compiler will not call the function where you have calls in the code, but rather will copy the code from the function to the spot where it is called. This is done to increase speed. Although this is a popular feature there is nothing in the C standard for inline functions. This means each compiler has its own way to define an inline function. For the CCS C compiler, the compiler will automatically tag some functions for inlining. To manually do this the **#inline** is used before the function and before the prototype if you have one. For example:

```
#inline
int  AddAndMul( int a, int b, int c ) {
        return (a+b)*c;
}
```

Nested Functions

A function declared inside a function is called function nesting and it is not legal in C. In C all functions in a compilation unit have the same global scope. The static qualifier may be used on a function declaration to hide it from other compilation units.

Recursive Functions

When a function calls another function that directly or indirectly ends up calling the original function, it is called recursion. Most C compilers allow this although care must be taken by the programmer not to get into an infinite loop. This is implemented using a data stack, where for each call a new space is allocated for the new local variables. Because the PIC® microcontrollers do not have a strong data stack, the CCS C compiler does not permit recursion. Larger computers are able to access data on a stack as easily as it is accessed from fixed memory. This is not the case for a PIC®.

A Little More on Sequence Points

Consider this function call:

```
f(x++,x++);
```

If **x** is 5, for example, what value is passed for the first and second parameter? It would depend in this case in what order the parameters are evaluated. The answer is that this is not valid code according to the C standard because there is a sequence point at each parenthesis. Most compilers allow this because they do not issue sequence point related errors. In short it is up to the compiler to decide what order arguments are evaluated in. Do not write code that depends on the order.

Well-Structured Programs

It is considered bad form to write your entire program in a single function called main(). For some programmers breaking a large project into functions comes naturally. For others, especially for large and/or complex projects, the program structure is not obvious. To help in breaking up a program into functions and to ensure a well-organized, maintainable, and verifiable result, there are various programming methodologies. Different methodologies work better for certain kinds of projects and certain programming languages. Some coding standards dictate a methodology.

One hot methodology you will hear a lot about is object-oriented programming. In short, this methodology defines all the objects in your program, then defines the operations and data for those objects. A diagram is created showing the relationship between objects. For example, objects might be a keyboard, database, and printer. This methodology meshes well with C++ because C++ has an element called a class that C does not have. A class is a group of functions and data items that are bundled together. Programmers can declare multiple copies of a class and can build new classes based on other classes. This fits in nicely with the object-oriented methodology. To make a class work we need more support from the microcomputer instruction set than we have in a PIC®. It is still possible to use this methodology with C without using a class, it is just not as elegant.

A very good traditional methodology for programming is called top-down. It is quite simple to organize any program using top-down. Start by looking at what the entire program needs to do and summarize it in three to seven topics like you would for an outline. Then for each of those items, further split it into the functions it needs to perform again at a high enough level so you have less than eight items. You keep doing this for each item.

What is happening is the first three to seven items form main(). Each other item is a function. In many cases the function simply calls a few other functions. Sometimes in a loop and sometimes conditionally. As the functions are created, the best route for data flow is established based on which functions need to know what. Stop breaking up items (functions) when the entire function can be coded in C on one screen (fewer than 25 lines). Again, some of these numbers may be specified in a coding standard.

At this time you have a call tree with main at the top, the functions main calls under it, and the functions they call under that. The tree was created top down (see Figure 7.4).

Further refinement of the code can be done by merging functions that are similar into a common function called in multiple places. The tree can then be further analyzed for data flow to ensure an effective structure.

There will always be cases when you should break the rules. The methodology is to help organize your program. Do not let it force you to make the code more difficult to understand. For example, consider a function that accepts a command from the user then calls one of 200 functions based on the command. It may be best to code this as a single function with 200 if statements or 200 cases. To break this up into 10 functions to meet the 25-line rule would make the code harder to read. The same is true when there are a large number of variables that need to be initialized.

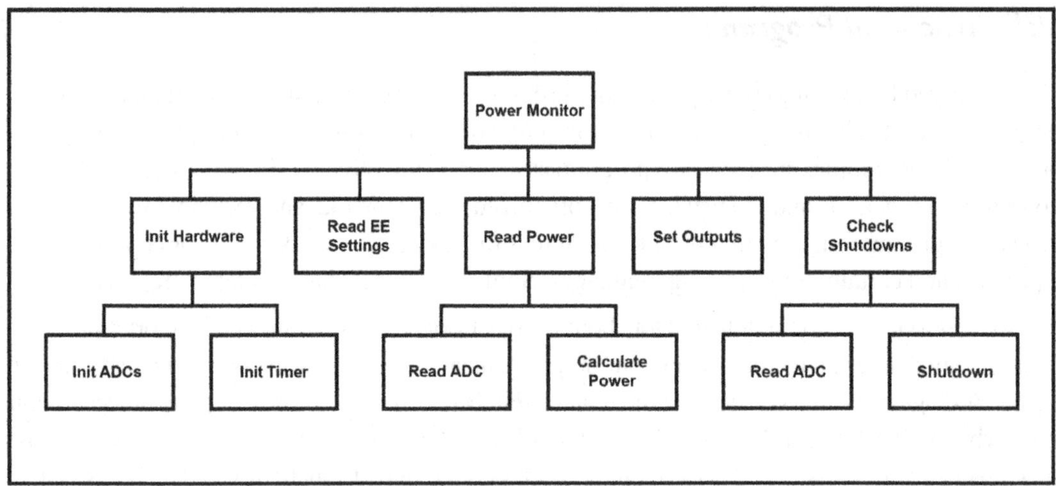

Figure 7.4: Example function call tree.

Given the above example it would also be bad practice to have the actual code for the 200 commands in the one function. Calling another function is one line, but to have several lines of code for each item would be a bad idea.

For larger programs it is best to proceed with testing bottom up to the extent it is possible. Simply calling a function with all the possible arguments and data sets is a great way to make sure it is performing as intended. Once tested then that function can be called by functions higher up on the tree when you test them.

Design Documentation

A lot can be understood about a program if the function structure is understood. The call tree is usually a key part of any design documentation. A full listing of functions and their purpose can also be a big help. It is common in the code to include a comment block at the start of each function to indicate what the function does and to identify the data passed in and out of the function. If the function uses global variables sometimes this will be identified here as well. Some coding standards will have a specific format to follow for this comment block.

There are also tools that can extract all these comments from a source code file and generate a document with a summary of the functions in the code. The CCS IDE has such a tool in the documentation ribbon.

Call trees can also be automatically generated from code. The following is an example of function comments. The compiler associates any comments directly before a function as belonging to the function for the purpose of automatic documentation generation.

```
////////////////////////////////////////////////////////////////////////////////
///
/// HostQueuePacket()
///
/// Queues a packet to be sent to the host.  This will format the packet
/// and place it in the transmit queue.  HostTask will then send send
/// the packet later.
///
/// Inputs:
///   cmd - Command to send
///   payload - Payload data to send
///   len - Lenth of the payload
///
/// Outputs:
///   BOOLEAN - Will return TRUE if the command was placed in the queue.
///             Will return FALSE if it could not be placed in the queue.
///
////////////////////////////////////////////////////////////////////////////////

BOOLEAN HostQueuePacket(int8 cmd,int8* payload,int len){
```

Implementation Details

Microprocessors have a special memory area called a stack. Think of the stack like an in-box on your desk. When you take an item out of the in-box it is the last item put in. The stack is primarily used to manage function calls. When a function is called, the address to return to is pushed on to the stack. When that function calls another function another return address is pushed on the stack. When a function does a return, the top address on the stack is popped off and that tells the CPU where to return to.

Traditional C compilers use the stack for more than return addresses. They will push arguments to the function on the stack as well as local variables. When a function terminates, the local variables and parameters are popped off the stack. This is a very elegant method of memory allocation that works well on most processors.

Unfortunately, using the stack for anything other than return addresses is not possible on most PIC® chips and not practical on the others. The compiler needs to work around this limitation and the fact that the stack maximum size on many PIC®s is small.

The CCS C compiler uses three methods to implement a function call and return:

1. The PIC® CALL/RETURN is used, requiring a single stack location.
2. If a function is only called once in the program then instead of a CALL a jump (GOTO) will be used to get to the function and the function will jump back to the instruction after the call. This saves valuable stack space.

3. If a function is marked #inline or if there is no more stack space left the compiler makes a copy to the called function at the point it is called. In this case there is no need for a CALL/RETURN or GOTO/GOTO.

Summary

- A function is a grouping of code that can be executed by other functions.
- The main() function starts when the program starts.
- Functions may have data passed in from the callers argument list to the functions parameter list.
- Data is traditionally passed by value (copied); however, compilers with a C++ extension allow a pass by reference (original item is used).
- Functions can return a single value as a result to the calling expression.
- Default parameters make some arguments optional.
- Function overloading allows functions with the same name and different parameter types or numbers.
- Functions cannot be defined inside functions.
- Recursive function calls are not allowed on the PIC.
- Top down is a sound methodology for organizing a program.

Exercise 7-1

Objective: Gain experience creating and using C functions.
Requires: E3 module, USB cable, PC.

Steps/Technical Procedure	Notes
1. Write a program to calculate the weight of water in a complex aquarium. The aquarium is made up of two cuboid areas connected by a tube. Both cuboid areas are the same size. The user should be prompted for the width, length, and height of the cuboid areas and for the diameter and length of the tube. All entries should be in feet and whole inches. To help in the design of this program, the following functions should appear in the code: get_size_data - Should return five numbers (in inches) calculate_weight - Five dimensions in, single weight out enter_feet_inches - Prompts user for feet and inches, returns total inches find_area_of_rectangle - Given length and width returns 2D area find_area_of_cuboid - Given 2D area and height returns area in inches^2	

find_area_of_circle - Given diameter returns 2D area find_area_of_tube - Given 2D area and length returns area in inches² area_to_weight - Given inches² returns pounds, uses the constant of 6.4279 pounds per cubic foot.	
2. The code should have no global variables and should loop around to keep asking for input and outputting the final weight. Functions input and output 32-bit integers for inches, square inches, and cubic inches. Internally the functions may use floating-point math.	
3. Modify the above program so each function also passes back an error flag. When an error is detected calculations should stop and the error should be sent to the user from only the calculate_weight() function. An error should be flagged when any calculation would result in a number too big to fit in the int32.	

Quiz

(1) Of the following statements, which one is true?
- (a) Each C file may have only one function
- (b) All functions must appear inside the main() function
- (c) All functions must pass back exactly one value
- (d) All functions must have a prototype before the function is first used
- (e) None are true

(2) A common C function feature that cannot be implemented on a PIC® is what?
- (a) Overloaded functions
- (b) Default parameters
- (c) Nested functions
- (d) Recursive functions
- (e) Reference parameters

(3) Given the following function and call, what is the value of x after the call?

```
int16 x;

int16  calculate( int8 x, int8 y ) {

   return  (long)x*y;

}

...

x=calculate(300,3);
```

(a) 900
(b) 123
(c) 765
(d) 132
(e) This code is not legal, there will be a compile error

(4) When is it OK to pass more function arguments than there are parameters?
(a) When there are default parameters
(b) When the function is overloaded
(c) Only for recursive functions
(d) Any function call, the extra arguments are a single expression combined with the comma operator
(e) Never

(5) What is wrong with the following function call?

```
funct1( funct2(3) );
```

(a) Recursive functions calls are not permitted on a PIC®
(b) Two functions cannot execute at the same time
(c) Function arguments must be numeric expressions that can be passed by value
(d) All of the above
(e) Nothing is wrong with this call

(6) Why are programs broken down into many functions?
(a) To make the code more readable
(b) To reduce the program size by putting duplicate algorithms into a function
(c) To reduce the scope of variables to just the areas they are needed
(d) All of the above
(e) None of these reasons

(7) Assume there is a stopwatch for every function and it is started when each function is entered and stopped when the function returns. When a function is re-called, the stopwatch continues where it last stopped. What are the times on the stopwatch for each function when the following code executes?

```
void functA(void) {
        delay_ms(10);
}
void functB(void) {
        delay_ms(20);
        functA();
}
```

```
void functC(void) {
          delay_ms(30);
          functA();
}
void functD(void) {
          functC();
          delay_ms(40);
          functB();
}
void main(void) {
          functD();
          functD();
}
```

(a) A = 40 ms B = 40 ms C = 60 ms D = 80 ms
(b) A = 10 ms B = 20 ms C = 30 ms D = 40 ms
(c) A = 40 ms B = 60 ms C = 80 ms D = 220 ms
(d) A = 20 ms B = 40 ms C = 60 ms D = 80 ms
(e) A = 20 ms B = 30 ms C = 40 ms D = 110 ms

(8) For the following function and function call, what will the value of x be?

```
int  calculate( int x, int y ) {
   if(x>5)
     x=5;
   return  x*y;
}
          ...
x=calculate(1*2*3, 4*5);
```

(a) 100
(b) 120
(c) 4
(d) 20
(e) None of the above, this is not legal C

(9) Which statement about the count of items in this code segment is true?

```
void  functA(int x, int y = 0);
void  functA(int x, int y = 0) {
}
    ...
functA(3);
```

(a) There are two functions
(b) There are two arguments
(c) There are two parameters
(d) There are two calls
(e) There are two statements

(10) For the following function and function call, what will the value of z be after the calls?

```
int  calculate( int & x, int y ) {
   x=x+1;
   y=y+2;
   return  x+y;
}
                 ...
int  a=1, b=2, z;
z=calculate(a, b);
z=calculate(a, b+z);
```

(a) 3
(b) 10
(c) 11
(d) 13
(e) None of the above, this is not legal C

CHAPTER 8

Arrays

An array is a list of values. Arrays have *elements* and all of the elements are the same data type. In math circles, arrays are sometimes called vectors.

An array declaration is the same as any other data declaration except it uses square brackets to indicate how many elements are in the array. Some examples:

```
int table[20];
char zipcode[10];
signed long x[3];
```

Arrays are accessed using the square bracket operator with an index. The index indicates what element is to be accessed. The first element is always index 0. For zipcode above, that means you can use index values from 0 to 9.

It is important to understand the `[ ]` used to access array elements are true operators and not part of the identifier or variable. That means you can follow an expression with `[ ]` and probably get into trouble. More on that in the Chapter 10. For now, do not do it.

Here are some examples of an array access:

```
average = ( data[0] + data[1] + data[2] ) / 3;
for(i=0; i<sizeof(table); i++)  sum+=table[i];
value = lookup[ base + offset ];
```

Array Initializers

Array initializers use braces to group the elements. For example:

```
int odds[5]={ 1,3,5,7,9 };
```

If there are not enough elements in the group for the array then the C rule is it fills the remaining elements with 0. For example:

```
int  data[10] = { 1,2,3 };  //  Gets 1,2,3,0,0,0,0,0,0
int  data[10] = {0};        // Gets all zeros
```

Embedded C Programming. http://dx.doi.org/10.1016/B978-0-12-801314-4.00008-9

Constant Arrays

Constant arrays, when used with the **const** or **rom** qualifiers in the CCS C compiler, are put into ROM. This is used for tables that need not change at run time. For the CCS C compiler, **const** is more efficient; however, you can only access the elements with the indexes that are shown in this chapter, not pointers as in Chapter 10. Here is an example:

```
const char hex[16] = {'0','1','2','3','4','5','6',
'7','8','9','A','B','C','D','E','F'};
```

Given the above declaration of a constant array, the following code will output a two-digit hex value for a given data byte. Review the code to understand how it works:

```
cout << hex[data>>4] << hex[data & 0xF];
```

String Variables

Constant strings have already been covered. These are the double-quoted strings that are always terminated with a 0. To declare a string variable we need the array construct. The elements of the array are **char**s. For example:

```
char mystring[12];
```

This will allow for a string of up to 11 characters followed by a 0 terminator. The standard C way to assign a constant string to a variable string is to use the **strcpy()** function like this:

```
strcpy(mystring, "Hello World");
```

The CCS C compiler also allows this shorthand notation:

```
mystring="Hello World";
```

To declare the variable with an initializer, use any of the following:

```
char  mystring[12] = "Hello World";
char  mystring[12] = {"Hello World"};
char  mystring[12] = {'H','e','l','l','o',' ','W','o','r','l','d',0};
```

There are many C standard functions to operate on strings. All assume the 0 terminator. These are covered in more detail in Chapter 12. For now, know the following are two ways to output a string:

```
printf("%s", mystring);
cout << mystring;
```

Dimensionless Arrays

A dimensionless array declaration looks like this:

```
int version[]="11.22.33";
```

The array has a dimension (in this case 9); however the declaration does not. With this syntax the compiler counts the elements in the initializer and uses that for the dimension. Another place you may see this is a parameter list like this:

```
int average(int list[], int count) {
```

In this case different size arrays could be passed into the function. The function may need to know the size of the array and that is why, when this is used, frequently a size of some sort is also passed in.

Finally you may also see a dimensionless declaration with no initializer and not in a parameter list. It looks like this:

```
int table[];
```

This is in fact not an array but a simple pointer. Since we are putting off a pointer discussion until Chapter 10 we will not be covering this syntax further here.

Multidimensional Arrays

So far our arrays have had one dimension. C allows for many dimensions on an array. If you look at a one-dimensional array as a simple list you can look at a two-dimensional array as a matrix. Figure 8.1 shows a simple example of one- and two-dimensional arrays:

For this example, the element table[1][0] is the value 4. The first dimension is thought of as the table row (vertical in our diagram) and the second as the column (horizontal in our table).

```
int   table[3]   =   {1,2,3};
```

1	2	3

```
int   table[2][3]   =   {{1,2,3},{4,5,6}};
```

1	2	3
4	5	6

Figure 8.1: One- and two-dimensional arrays.

Note the extra `{ }` in the initializer for each row. This helps to see how the data is placed in the array, especially for large arrays. It is, however, allowable to initialize like this:

```
int table[2][3]={1,2,3,4,5,6};
```

Although more dimensions than two are allowed, memory is used up very fast. A $10 \times 10 \times 10$ array is 1000 elements and even as an int8 that is quite a bit of memory for a PIC®. In a multi-dimensional array, only the last dimension is allowed to be dimensionless. This is not legal:

```
int table[][];
```

But the following is legal:

```
int table[5][];
```

Index Range

Using an improper range of an *array index* is a common problem in C programs and it can lead to very difficult to find problems. For example, the array declared like this:

```
int list[20];
```

should have index values from 0 to 19. It is a common problem where an index of 20 is used. If this is done there will be a write to a memory location not part of the array, and that will cause unpredictable program operation.

C requires all elements of an array to be saved in consecutive memory locations. Some PIC® chips have RAM split up so there are only a limited number of consecutive locations. For example, some parts have at most 31 bytes together and many of the popular PIC16 parts have at most 96 bytes in consecutive locations. This may require you to split an array up into smaller arrays to get all the data to fit in a PIC®.

Example Array Usage

The following is an example function that adds each corresponding array element together and saves the result in the first array.

```
void  Vector_Add( int vectorA[], int vectorB[], int count ) {
      for( int i = 0; i<count; i++ )
         vectorA[i]+=vectorB[i];
}
```

Because we do not know the array size it is passed in as a parameter. The following is an example call:

```
Vector_Add(A, B, sizeof(A));
```

The above only works if the array element is a byte. A more general call that will work for any type looks like this:

```
Vector_Add(A, B, sizeof(A)/sizeof(a[0]));
```

In computer programs, it is a common practice to sum up a series of numbers to create a checksum. This checksum can then be kept with the list of numbers and later tested to ensure a number did not change. This might be used for a data transmission or if data is stored on some device. The following is an example function to calculate a checksum:

```
int   checksum( int data[], int count ) {
      int result=0;
      for( int i = 0; i<count; i++ )
         result+=data[i];
      return(result);
}
```

It is interesting to note that the above function could be used with a two-dimensional array as well. Because in C the data for a two-dimensional array is saved in memory with each row after each other, the following would work. A 5×3 array is saved in memory like a 15-element array.

```
int table[5][3];
cs = checksum( table, sizeof(table) );
```

Here is a function that counts the number of words in a string:

```
int   words( char string[] ) {
      int result=0, i=0;
      while(data[i]!=0)
         if(data[i++]==' ')
              result++;
      if(data[0]!=0)
         result++;
      return(result);
}
```

Use of the ++ operator inside a subscript is a common practice to iterate an index through the elements of an array. This code is counting spaces. The number of words is assumed to be the number of spaces plus one, unless the string is empty. That is what the last *if statement* is checking for.

Lookup Tables

A lookup table (array in C) may be used to convert one number to another. One option is to convert numbers ranging from 0 to 9 to seven-segment signals to drive a display (see Figure 8.2).

The proper seven-segment code may be pulled from an array of constants by adding an offset to the array index. The offset is the number we want to display. The seven-segment binary code for the number is stored as that array element, as shown in Figure 8.3.

For demonstration purposes, we will set the digit to 2. The seven-segment equivalent bit pattern will be accessed in the table/array and "2" will be displayed.

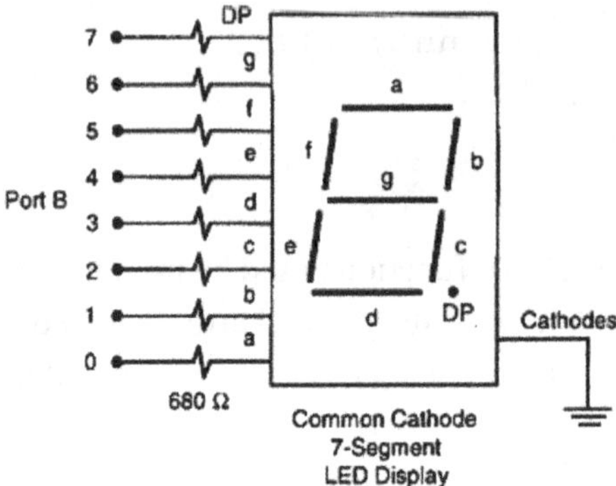

Figure 8.2: Common cathode 7-segment LED schematic.

Number	7-SEGMENT Code	DPgfe dcba
0	0x3F	0011 1111
1	06	0000 0110
2	5B	0101 1011
3	4F	0100 1111
4	66	0110 0110
5	6D	0110 1101
6	7D	0111 1101
7	07	0000 0111
8	7F	0111 1111
9	6F	0110 1111

Figure 8.3: Digit to hex pattern translation table.

```
main()
   {
      byte const nums[10]= {0x3f,0x06,0x5b,0x4f,0x66,0x6d,0x7d,0x07,0x7f,0x6f};


      int digit;

      digit = 2;
      output_b (nums[digit]);


      while (TRUE);
   }
```

Searching Arrays

The simple approach to search through an array is to start at the beginning and increment through the array until the item is found:

```
for( i=0; i<num && data[i]!=key; i++) ;
if(i<num)
   found=i;
else
   found=-1;
```

This is called a linear search and it is slow. Another popular algorithm is a binary search. For this to work the data must be sorted. Figure 8.4 is an illustration to help understand the algorithm.

A binary search starts in the center and checks to see if the item we want is above or below. We then look halfway between the pivot point and the opposite end of the array. The item will be found very quickly. The following is what the code looks like. Use Figure 8.4 to follow through the algorithm. Spend some time reviewing it to get used to visualizing algorithms from the code.

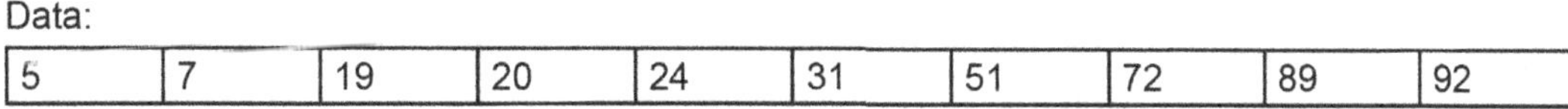

Data:
5
7
19
20
24
31
51
72
89
92

Figure 8.4: Sorted data array illustration.

```
// Before we get here the data is in array data
// num is the number of elements
// key is what we are searching for

   int p,q,pivot;
   signed int found;

   found = -1;
   p = 0;
   n = num;

   while (n > 0)  {
      pivot = n >> 1;
      q = p + pivot;

      if (key < data[q])
         n = pivot;
      else if (val == 0) {
         found = q;
         break;
      } else {
         p = q + 1;
         n -= pivot + 1;
      }
    }
```

What if the data is not sorted and cannot be because the position is important? What we do is create a second array with sorted indexes. Figure 8.5 shows an example with a result. 3 is the index of the lowest value, 2 is the next sorted index, then 8, and so on.

In the search algorithm, the **`if`** statement now looks like this:

```
if (key<data[sorted[q]])
```

And when found we do this:

```
found=sorted[q];
```

Frequently searches will involve not a simple array like this but rather an array of structures. Structures are covered in Chapter 9.

Data:

89	20	7	5	51	24	92	31	19	72

Sorted:

3	2	8	1	5	7	4	9	0	6

Figure 8.5: Sorted data array illustration.

Sorting Arrays

The binary search needs your array to be sorted and there are many other situations where data needs to be sorted. The slow and crude method is called an insertion sort. As each item needs to be inserted into the array you find the right spot and shove everything else up to make room. If the array already exists you find the smallest value, trade it with whatever is in position 0, then do the same starting at position 1 and going up. There are a number of more advanced algorithms although it can be difficult to figure out how they work. The following Shell algorithm (developed by Donald Shell) has excellent performance.

```
unsigned int m,j,i,l,t;
int1 done;

m = num>>1;
while( m > 0 ) {
   for(j=0; j<(num-m); ++j) {
      i = j;
      do {
         done=TRUE;
         l = i+m;
         if( data[i]>data[l] ) {
            t=data[i];
            data[i]=data[l];
            data[l]=t;
            if(m <= i)
               i -= m;
            done = FALSE;
         }
      } while(!done);
   }
   m = m>>1;
}
```

Data:

9	4	4	5	1	2	2	1	9	7

Could be easily and quickly expressed like this:

0	2	2	0	2	1	0	1	0	2

Figure 8.6: Illustration of sorting by counting.

Again, sorting is most often done on structures. If, however, you do need to sort an integer array and the range is limited, consider simply counting each value. An example is shown in Figure 8.6.

Summary

- Arrays are a list of data items all of the same type.
- Array elements are referenced using an index from 0 to the array size minus one.
- Arrays may be initialized using the `{ }` grouping symbols.
- The `[ ]` notation is used when the compiler is to count the array elements in an initializer or when an array is passed to a function with no specific size.
- Each array subscript is a dimension to the array and arrays may have multiple dimensions.
- String arrays must allocate an extra location for the 0 terminator.
- The Shell sort is a very efficient method to sort an array.
- The binary search is a very fast way to search for data in a sorted array.

Exercise 8-1

Objective: Gain a working knowledge of using C arrays.
Requires: E3 module, USB cable, PC.

Steps / Technical Procedure	Notes
1. Write a program that enters 10 numbers from the user into an array. After the 10 numbers are entered output them to the screen in reverse order.	
2. Write a program that allows the entry of a series of numbers from 1 to 100. Entry stops when a 0 is entered. After the 0 is entered display to the screen how many of each number were entered. Skip output of numbers that had no entries. For example, if the numbers entered are 20,30,1,20,1,0, then the output is: 1=>2, 20=>2, 30=>1.	
3. Using the words() function in this chapter write a program to enter a string of up to 100 characters and output the number of words.	

4. Modify the above program so that it works if there are any number of spaces between words and to allow for spaces before the first word or after the last word.	
5. Write a program to allow entry of numbers (single bytes). After each number is entered output to the screen the average of the last five entries. This is a simple filtering algorithm.	

Quiz

(1) For the following array, what are the number of subscripts, number of elements, and highest index?

```
long i[10];
```

(a) 10, 9, 9
(b) 9, 20, 10
(c) 0, 9, 10
(d) 1, 10, 9
(e) This declaration is not valid

(2) What will the following code display?

```
int8  table[10] = {1,2};
for( int sum=0, int i=0;
   i<sizeof(table);i++)
      sum+=table[i];
cout << sum;
```

(a) 1
(b) 3
(c) 15
(d) 21
(e) 255

(3) For the following declaration, how many bytes of RAM will be allocated?

```
Int32 table[5][5];
```

(a) 0
(b) 5
(c) 20
(d) 25
(e) 100

(4) Assuming the following declaration, what would the value of data[4] be?

```
char data[] = "5678";
```

(a) 0
(b) 'C'
(c) 4
(d) '8'
(e) This is not legal

(5) What is the output from the following code?

```
int  table[] = {1,2,3,4,5,6,7,8,9};
for(int i=1; i<=3; i++)
   cout << table[ i*3 % 9];
```

(a) 369
(b) 470
(c) 240
(d) 18241
(e) This is not valid C

(6) Which of the following algorithms are not a good use for an array?
(a) Maintain the average of the last 50 samples
(b) Hold a group of data points for processing
(c) Wait for a input value over a predefined threshold
(d) Keep counts of specific numbers
(e) A lookup table of cos values in floating-point format

(7) What does the following code do?

```
for (i=0; s1[i]!=0; i++) {
   for (j=0; s2[j]!=0; j++)
      if( s1[i+j]!=s2[j] )
         break;
      if(s2[j]==0)
         cout << i;
}
```

(a) Outputs the number of times the s2 string appears in the s1 string
(b) Outputs the positions the s2 string appears in the s1 string
(c) Outputs the positions the s2 string does not appear in the s1 string
(d) This code does nothing useful
(e) This is not legal C

(8) The following code is an attempt to transverse a matrix (switch rows and columns). What is the flaw in this code?

```
for(i=0; i<sizeof(matrix[0]); i++)
   for(j=0; j<(sizeof(matrix)/sizeof(matrix[0]); j++)
      matrix[j][i] = matrix[i][j];
```

(a) Matrix has only one subscript in the sizeof
(b) The **<** operators should be **<=**
(c) The loops should be sequential not nested
(d) Data not yet moved will be overwritten
(e) No flaw, it will transverse any two-dimensional matrix

(9) What is the problem with the following declaration?

```
unsigned int8 mystring[10]="ABCDEFGHIJ";
```

(a) Characters cannot be unsigned
(b) Cannot use an int8 for a character string
(c) Cannot have an array size and an initializer
(d) The count for the array size is wrong
(e) All of the above

(10) With the following encryption algorithm, what will the value of encrypted be after processing?

```
char key[] = "QWERTYUIOPASDFGHJKLZXCVBNM";
char plaintext = "HELLO";
char encrypted[10];
for( i=0; plaintext[i]!=0; i++)
   encrypted[i]=key[ plaintext[i] - 'A' ];
encrypted[i]=0;
```

(a) ITSSG
(b) QWERT
(c) OLLEH
(d) LCGGS
(e) SGGCL

CHAPTER 9

Structures

Structures are similar to arrays. Arrays have *elements*, and all of the elements are the same data type. Structures have *members* (sometimes called *fields*). Members may be all of one data type, or a mixture of data types as we shall see by way of examples. A structure is a way to group data that is related. Think of a record in a database program.

Name and phone number
Customer, address, phone, accounts receivable balance
Etc.

First, a structure type must be defined. Think of the definition as a layout, or plan, or template for the structure type you are creating. An instance of a structure is a structure variable which contains members that are also variables. A structure definition looks like:

```
struct structure-tag

{
  member definition;
  member definition;
  .

  .
  member definition;
}variable-identifier;
```

The tag and variable identifier are optional; however, at least one must usually be used. The tag is a name for this structure layout. This name may be used elsewhere in the code so the layout does not need to be repeated. The identifier is the regular name for this variable.

Here is an example:

```
struct   {
   int16 to;
   int16 from;
   int8  data[32];
   int8  checksum;
}  message_in, message_out;
```

Embedded C Programming. http://dx.doi.org/10.1016/B978-0-12-801314-4.00009-0

This defined two variables (message_in, message_out) that are structures. Each variable is 37 bytes in memory. To access the "from" in the message_out the syntax is message_out.from.

Here is some code to calculate the checksum:

```
for(message_out.checksum=0, i=0; i<sizeof(message_out.data); i++)
   message_out.checksum+=message_out.data[i];
```

This is the same definition using tags:

```
struct   msg_struct {
   int16 to;
   int16 from;
   int8  data[32];
   int8  checksum;
} ;
struct  msg_struct  message_in, message out;
```

Like the enum, the typedef may be used to eliminate the use of struct. Here is a typical typedef:

```
typedef struct  {
   int16 to;
   int16 from;
   int8  data[32];
   int8  checksum;
   }  msg_type;
msg_type  message_in, message out;
```

A structure can be referenced as a group or an individual member can be referenced. The member referencing is done with the operator.

Structures can be passed in whole to a function; however, care should be taken because a structure can take a lot of RAM. Since function parameters are pass-by-value, a full copy of the structure will be made.

Structure Nesting and Arrays

The members in a structure may be any normal data definition including another structure. Structures can be nested inside other structures.

It is also allowed to have an array of structures. Here is an example of both a nested structure and an array of structures:

```
struct   {
   struct {
      int8 node_id;
      int8 unit_id;
   }  to, from;
   int8  data[32];
   int8  checksum;
}  message_queue[10];
```

To access the unit_id for the from member of the first message in the queue, the reference looks like:

```
message_queue[0].from.unit_id
```

By now it should be clear how powerful structures can be for organizing data and how quickly the syntax can become complex.

Structure Layout in Memory

The members in a structure are located in memory one after another with the first member in the lowest memory location. This is all C guarantees however. Some compilers will insert space between structure members to satisfy alignment requirements of the processor. For example, the PIC24 chip can only access 16-bit integers in memory on an even-numbered address. In a structure with an int8 followed by an int16, the compiler may insert an unused int8 between the two. This can be prevented by using the packed qualifier on the structure. Using packed will prevent any gaps; however, there could also be some implementation problems so care should be used.

Consider the following structure. As is with the packed directive, on a PIC24 it will occupy 6 bytes. Without packed it takes 8 bytes, but that size is not guaranteed and may vary.

```
struct  _packed {
   int8  id;
   int16 length;
   int16 addr;
   int8  next_id;
};
```

There is no standard for the packed qualifier. Some popular compilers use the nonstandard attribute qualifier to do the same thing. The CCS C compiler also allows this alternate syntax:

```
struct __attribute__((packed)) {
   int8  id;
   int16 length;
   int16 addr;
   int8  next_id;
};
```

Bit Fields

The members in structures can be of any data type, plus there is one bonus type qualifier that can be used only in structure members. It is the bit field designator. An integer member can be identified to have exactly a specified number of bits. The number of bits can be up to the number of bits in an int (not more). Consider this example:

```
struct   {
   int8  x,y;
   int  red : 2;
   int  green : 2;
   int  blue : 2;
} point;
```

The red member may have the values 0,1,2,3 and it is exactly 2 bits in size. The whole structure is 3 bytes and in it 2 bits are not used.

One use for bit fields is to exactly match a specific data layout for communications messages. To save storage and transmission time, usually data is defined to the exact bit size needed. For example:

```
struct   {
   int  mac_mode :2;           // 0=dynamic,1=base fixed,2=mobile fixed,3=reserved
   int  MCS : 4;               // 0=unknown, 1-11 modulation coding methods 1-11
   int  mod_threshold : 1;     // 0=none, 1=fixed
   int  half_band : 1;         // 0=mobile, 1=base
   int  msg_type : 6;          // Types 1-31 or 0 for test
   int  reserved : 2;          // Always 0
} message_header_one;
```

Another use is to make a structure exactly match the layout of an internal processor register. The variable can then be placed at the register address and the members of the structure when accessed are accessing the hardware bits directly. For example, many PIC® processors have a register called SSPCON for the synchronous serial bus. The following is a sample definition:

```
struct {
   unsigned int SSPM0:4;
   unsigned int CKP:1;
   unsigned int SSPEN:1;
   unsigned int SSPOV:1;
   unsigned int WCOL:1;
} SSPCON;
#locate SSPCON = getenv("sfr:SSPCON")
```

This shows a field of 4 bits called SSPM0 followed by four 1-bit fields. To turn on the SSP module by writing a 1 to the enable bit, you need only do:

```
SSPCON.SSPEN=1;
```

The **#locate** is used to place the structure at a specific address. Instead of hard coding the address here we made a call to a CCS C compiler extension that will find the address for specific PIC® registers.

Unions

The cousin to structures is the union. The union looks like a structure except the members instead of appearing sequentially in memory are placed on top of each other. A simple and maybe useless example:

```
union {
   int8 a;
   int8 b;
   int8 c;
} x;
```

If one does a x.a=5 then any reference to x.b or x.c will show a 5. In fact x.a, x.b, and x.c are all put at the same memory address. This union takes up only 1 byte of RAM. Here is a more interesting union:

```
union {
   struct {
      int low : 4;
      int high : 4;
   } asnibble;
   int8  asbyte;
} x;
```

In this case the following code will output a 4 followed by a 5:

```
x.asbyte = 0x45;
cout << x.asnibble.high;
cout << x.asnibble.low;
```

A neat trick but not usually worth the effort. You will see them in cases where a data item can be interpreted in different ways, usually based on some field in the data. For example, consider a message packet that is sent to control a device. There may be three different kinds of messages but the receiver does not know ahead of time what kind of message is coming in. They are all 16 bytes. The data type may look like this:

```
struct {
   char  message_type;
   int16  from;
   int16  to;
   union {
      int32  id;                  // For message_type= 'I'
      struct {                    // For message_type 'M'
         int32  pos_x, pos_y;
         int8   direction;
      };
      struct {                    // For message_type 'E'
         int16 power;
         int8  mode;
      };
   };
   int16  checksum;
}message;
```

Notice all the members of the union are not the same size. The size of a union is the size of its largest member. Another C feature to note is the inside union and structs do not have a member name. The name is optional for structures and unions inside a structure or union. When the name is not used the member operator is skipped. For example, **`message.power`** is a legal reference. If would not be a legal reference if either the union or struct had an assigned name.

Example of Structures in a Program

```
typedef signed int8 temperature;  // Degrees Fahrenheit
typedef int32 time_type;          // milliseconds since midnight
typedef int32 date_type;          // Days since Jan 1, 2000

struct datetime{
   time_type  time;
   date_type  date;
};

#define MAX_SENSORS 10

typedef
   struct {
      struct datetime last_sample;
      int16    sample_count;
      int8     sensor_count;
      struct {
         temperature last;
         temperature min;
         temperature max;
      } sensor[MAX_SENSORS];
   } SENSORS_TYPE;


void read_temperatures(SENSORS_TYPE & sensors) {
   sensors.sample_count++;
   get_datetime(sensors.last_sample);
   for(int i=0; i<sensors.sensor_count; i++) {
      sensors.sensor[i].last = read_temperature(i);
      if(sensors.sensor[i].last>sensors.sensor[i].min)
         sensors.sensor[i].min=sensors.sensor[i].last;
      if(sensors.sensor[i].last>sensors.sensor[i].max)
         sensors.sensor[i].max=sensors.sensor[i].last;
   }
}

void main(void) {
    SENSORS_TYPE sensors;
    do {
       read_temperatures(sensors);
       display_data(sensors);
    } while(TRUE);
}
```

Summary

- Structures are data types with multiple members, each with its own type.
- Structures may have tags like the enum.
- Structures are be referenced in whole or by individual member.
- Structures may have structures nested within.
- Integer structure members allow the specific number of bits for the member to be specified.
- Unions are like structures except that all members appear on top of one another in memory.

Exercise 9-1

Objective: Learn how to use C structures in common microcontroller applications.
Requires: E3 module, USB cable, PC.

Steps / Technical Procedure	Notes
1. Write a program to that prompts the user for an entry or search. If the user enters an E, then in the entry mode ask for a part number (int16), a row number (int8), a shelf ID (A-Z), and a bin number (int16). Save this data in the next available entry in an array of structures. If an S is entered, then prompt for a part number and display the row, shelf, and bin.	
2. Microprocessors have an internal register called the status register. This register among other things has flags to indicate what happened in the last math operation. One bit in that register is the carry flag. The carry flag is set to 1 if you add two numbers and the size exceeds the number of bits in the number. For example, an int8 add of 200+100 will cause the carry flag to be set to 1. On the PIC??? the layout for the status register is as follows: Carry flag — One bit Decimal carry flag — One bit Zero flag — One bit Overflow flag — One bit Negative flag — One bit Reserved — Three bits Write a program to define a structure for the status register and place it on top of the PIC® status register so these flags can be accessed from C. Construct your program to prompt the user for two numbers. Add the numbers and save the carry flag to an int1 variable. Then output to the user the status of the carry flag by outputting the variable. You are not able to output the carry flag directly in the cout because other math operations may be done in the processor before the flag is read. You want to read the real flag as soon after the math operation you are testing as possible.	

3. Create a union of a floating-point number and a 4-byte array. Construct the program such that the user enters a floating-point number and the program outputs the value of each of the 4 bytes that make up the floating-point number in hex. The cout can be used to output a hex number like this: `cout << hex << value;` In the IDE, use the TOOLS > NUMERIC CONVERTER to see if the answers are right.	

Quiz

(1) In the following code, what is wrong?

```
struct  s1  {
   int x;
}  s2;
struct  s1  s3;
s1 s4;
```

(a) You cannot define the structure with both s1 and s2
(b) You must have more than one member in a structure
(c) You cannot use struct when defining s3
(d) You must use struct when defining s4
(e) Nothing is wrong

(2) In the following code, what will the value of x be?

```
int16 x;
struct  {
   int8 a;
   int8 b;
}  s;

s.a=0x12;
s.b=0x34;
x = s;
```

(a) 0x12
(b) 0x34
(c) 0x1234
(d) 0x3412
(e) This is not legal C

(3) How many bytes of RAM are allocated with the following declaration?

```
struct  s1 {
   int8 a,b;
} s2,s3;
```

(a) 0
(b) 1
(c) 2
(d) 4
(e) 6

(4) How many bytes of RAM are allocated with the following declaration?

```
union  s1 {
   int8 a,b;
} s2,s3
```

(a) 0
(b) 1
(c) 2
(d) 3
(e) 4

(5) What is wrong with the following code segment?

```
struct  {
   int8 a,b;
   int8 c[10];
   } s1,s2;
...
s1.a=s2.c[s1.b];
```

(a) You can have arrays of structures but not arrays inside structures
(b) You cannot index an array with a structure member
(c) You cannot have the same structure on both sides of the assignment operator
(d) Structure array subscripts must be constant
(e) Nothing is wrong with this code

(6) What is the problem with creating a union to dissect the elements of a floating-point number like this?

```
union {
   float  asfloat :32;
   struct {
   int  sign : 1;
   int8 exponent;
   int  mantissa:23;
}
```

(a) You can not specify the number of bits in a float like this
(b) Without a packed qualifier there will be a gap between sign and exponent
(c) The size of the mantissa is larger than an int
(d) All of the above
(e) There is nothing wrong with this declaration

(7) What is output from the following code segment?

```
union {
   struct {
      int  a : 3;
      int b : 3;
      int c : 2;
   } aa;
      int8 bb;
} u;
int16 x;
u.aa.a=0;
u.aa.b=6;
u.aa.c=1;
x = u.bb;
cout << hex << x;
```

(a) 70
(b) 112
(c) 61
(d) 38
(e) 160

(8) Which of the following statements is true?
 (a) Structure members cannot have the same name as the structure
 (b) Structures inside structures cannot have the same name as other members in the outer structure
 (c) Two structures inside a structure cannot have member names that are the same
 (d) All are true
 (e) None are true

(9) What does the following code output?

```
union {
    struct {
        int1 a,b,c,d,e,f,g,h;
    }bb;
}cc:
cc.aa=-12;
cout << cc.bb.h;
```

 (a) 0
 (b) 1
 (c) 12
 (d) −12
 (e) −1

(10) For the structure in question 9, which of the following structure accesses is invalid?
 (a) cc.bb.a=cc.bb.b
 (b) cc.aa=cc.bb.c
 (c) cc.bb.d=cc.aa
 (d) cc.aa=cc.bb
 (e) All are invalid

CHAPTER 10

Memory and Pointers

This is a very important chapter to fully understand if you expect to become expert at C. The topics in this chapter are often misunderstood and oftentimes incorrectly used, producing the most difficult of problems to solve.

Memory

A processor accesses memory using a numerical address. The memory contents have a specific size for all addresses. We will start by only discussing RAM and then add ROM later in this chapter. Each example will assume each address in RAM has an 8-bit data item. This is true for all PIC® processors.

Numerically, addresses start at 0; however, the actual locations that have usable RAM may be limited. For example, usable RAM on the PIC16F887 chip starts at address 0x20. The addresses from 0x00 to 0x1F reference special registers in the part that control the processor and peripherals called special function registers (SFRs).

The C compiler will allocate program variables to specific memory locations. For example, the following is a visual representation of the memory allocation for a handful of variables:

```
int  a = 5;
int  b = 35;
long c = 1234;
float  f = 1.23;
int  x[5] = {5,4,3,2,1};
```

Using Figure 10.1, we can then say the B variable is located at address 0x21 and has a value of 0x13. The C variable is located at addresses 0x22 and 0x23.

The CCS compiler has an *output file* (the .SYM file) that will show how memory for a program is allocated. The following is an example snip from a .SYM file:

Embedded C Programming. http://dx.doi.org/10.1016/B978-0-12-801314-4.00010-7

20	05	a
21	23	b
22	D2	c
23	04	
24	7F	f
25	1D	
26	70	
27	A4	
28	05	x[0]
29	04	x[1]
2A	03	x[2]
2B	02	x[3]
2C	01	x[4]
2D		
2E		

Figure 10.1: Example RAM layout.

```
090-091 usb_put_packet.len
092     usb_put_packet.tgl
093-094 usb_put_packet.j
095     usb_put_packet.i
096-097 usb_put_packet.buff_add
098     usb_flush_in.endpoint
098     usb_put_packet.@SCRATCH1
099-09A usb_flush_in.len
```

The first line shows a variable named len in the function **usb_put_packet()** is located at addresses 0x90 and 0x91. Notice we have two variables at location 0x98. This is allowed because the two functions **usb_flush_in()** and **usb_put_packet()** are never active at the same time. Notice as well the variable that starts with a **@**. These are compiler-generated variables. In this case the compiler needed a scratch location to save some intermediate result. For example, in the evaluation of **(A+B) * (C+D)** the compiler must evaluate **A+B**, then save that value and calculate **C+D**, and finally multiply the result with the saved value. That saved value needs a scratch location. The .SYM file can be easily viewed in the IDE by clicking on the SYMBOL MAP icon on the compile ribbon. You can also view the addresses of the special function registers for your chip by clicking on VIEW > SFR's.

Address-of Operator

The unary operator **&** is the *address-of operator*. Assume the above memory allocation. The following code will output 20:

```
b = &a;
cout << hex << b;
```

Note that unlike most operators the **&** only works with an lvalue operand (usually a variable). An expression or constant is not legal.

Indirection Operator

The unary operator ***** is the *indirection operator*. The ***** will return the value at the indicated memory location. It is not a direct memory access but rather an indirect access. For example, the following will output 4 using the above memory map:

```
a = *b;
cout << hex << a;
```

This will output 1D:

```
a = *0x25;
cout << hex << a;
```

An expression that starts with an indirection operator may be used as an lvalue. For example:

```
*b = a;    // writes 5 to memory location 0x23
```

Forcing a Variable Address

Normally variables are assigned addresses by the compiler. Sometimes in embedded programming it is useful for the programmer to locate a variable at a certain address. This might be done to put a variable at the same address as a processor register so accessing the register is as easy as accessing a variable. It might also be done to force certain variables on top of one another as an alternative to the union. Some chips have memory that is faster to access than other memory. In this case one might want to put a heavily used data structure in the best memory. Using the CCS C compiler this is done with **#locate**. The following is an example of putting an array at location 0x20–0x2F:

```
int8  a[16];
#locate  a = 0x20
```

Pointer Types

In the **a=*b** from the above example, the **b** variable is referred to as a pointer. It points to another memory location. Although you can declare pointers like a regular variable, because of automatic type conversion you should declare the variable specifically as a pointer so the compiler can properly deal with the type matching and correct variable size. The proper way to declare **b** would be:

```
int * b;
```

This is a pointer to an int named **b**. The **(int *)** is the type. The ***** in this context is a pointer symbol. Although the variable **b** points to an **int** (one byte) the **b** variable itself is two bytes for most PIC® chips. That is because the memory map usually has more than 255 addresses.

Here are some example expressions and their types:

```
int  a,b;
long c;
float  f;
int  x[5];
int * d;
float * e;

&a    int *
&f    float *
&x[2] int *
&x    illegal
*d    int
*f    float
*c    int        (not recomended)
*x    int        (same as x[0])
```

Pointer Math

The **+, -, ++,** and **--** operators work in a special manner when one operand is a pointer. With the above declarations, the following statement:

```
e = e + 1;
```

will increment e by 4 not by 1. The pointer is incremented such that it points to the next object of the same type. Mathematically, think of it as:

```
pointer + constant * sizeof( *pointer )
```

Back to Subscripts

Earlier we pointed out the [ ] for array subscripts were actually operators. The formal definition is as follows:

`expr[ index ]` is evaluated as: `*(expr+index)`

This works because array names are actually pointers and can be used like any other pointer. Likewise subscripts can be applied to a pointer. Here is an example:

```
int  table[10];
int * p;

p = &table[5];
p[3] = 4;
```

This writes a 4 into table[8].

The array and pointers are so close that the following two declarations are identical:

```
int  *  x;
int  x[];
```

Be aware that although array names are pointers, structure names are not.

Back to Function Parameters

The original C specification did not provide for reference parameters and still many C compilers do not support reference parameters such as are shown in the first function below. The traditional way to pass back more than one value to a caller using only a *pass-by-value parameter* is to pass a pointer to the object you want the function to change. The function can then use the pointer to modify the original object. See the second function below for the traditional way to do the same thing.

```
void Find_Min_Max( int list[], int count, int & min, int & max ) {
    for( int i=0, min=255, max=0; i<count; i++) {
        if( min>list[i] )
           min=list[i];
        if( max<list[i] )
           max=list[i];
    }
}
// call like this:  Find_Min_Max( table, sizeof(table), lowest, highest );

void Find_Min_Max( int list[], int count, int * min, int * max ) {
    for( int i=0, min=255, max=0; i<count; i++) {
        if( *min>list[i] )
           *min=list[i];
```

```
        if( *max<list[i] )
           *max=list[i];
    }
}
// call like this:  Find_Min_Max( table, sizeof(table), &lowest, &highest);
```

Back to Structures

Because structures take a lot of space, pointers to structures are popular. It is more effective to pass a pointer to a structure than the structure contents. The syntax for accessing a structure member from a pointer is a little scary. Consider this example:

```
struct st {
   int  a,b;
}  sv;
struct st * sp;
...
sp = &sv;
(*sp).a = 0;
```

Here we have a structure tag **st**, a variable in memory **sv**, and we defined a pointer to a structure **sp**. The assignment is simple enough. Notice the **&** operator is needed to find the structure address (pointer). This is not the same as it was for arrays where the array name is a pointer. To access the member we needed to use the indirection operator *****; however, because the member operator **.** takes precedence over the ***** operator we needed to add the parens.

Because the above syntax is so popular, C has a special member operator for just this circumstance. The **->** operator is used as follows:

```
sp->a = 0;
```

This is most often used when passing a structure to a function. Just the pointer is passed and the **->** operator is used to access the members.

There is one more feature in C concerning pointers to structures. C allows a forward reference to a structure not yet defined by a member of a structure if it is a pointer. For example:

```
typedef  struct {
   int a,b;
   stype * next;
} stype;
```

In this case we have a structure definition that includes as a member a pointer to another structure of the same type as the structure being defined. One use of this is to create a linked list of structures. Each structure has a pointer to the next structure in the list. Linked lists are sometimes used instead of an array to allow for more flexibility in the number and order of the elements.

Function Pointers

This section is not about the function parameters but rather the function itself. C allows you to define a variable that is a pointer to a function. That variable can then be used to call the function that it points to. This is used in cases where you have code that might call one of a number of different functions inside the code.

Consider a function that performs some math operation and in order to report an error uses a function supplied by the caller. This allows the caller to decide what to do for errors. For example, increment an error counter, ignore it, light a red LED, or display an error on an LCD screen. To make the syntax easier to follow we will use a typedef to define the function pointer type. Here is the code including the call:

```
typedef void  (*error_handler_ptr)(int error_number);

int16 do_math( int8 a, int8 b, error_handler_ptr error_handler) {
   if(a>300) {
      (*error_handler)(1);
      result=0;
   }
   if(b==0) {
      (*error_handler)(2);
      result=0;
   }
      result = a*150/(b*3);
}
void show_error(int errn) {
   cout << "Error#" << errn << endl;
}
      ....
value = do_math(scale, units, show_error);
```

The typedef looks odd; it is similar to the call but the type name is where the function name would be. The second set of parens is the parameter list. The call uses the * operator and needs to be in parens due to precedence rules. Notice we do not use the & with the function name to get the pointer, simply the function name with no parens after it gets you the pointer.

Other Uses for Pointers

Pointers are often used to move through an array instead of using an index when you know you are only going through sequentially. For example, here is a function to count the spaces in a string:

```
int  count_spaces( char * ptr ) {
   count=0;
   while( *ptr!=0 )
      if( *(ptr++)==' ')
         count++;
}
...
char test[40] = "Hello to the world";
int n;
n = count_spaces(test);
```

Pointers can be assigned to other pointers and comparison operators can be used on pointers. In addition, a function can return a pointer. Here is a function that returns a pointer to the first space in a string:

```
char *  find_space( char * ptr ) {
    while( *ptr!=0 )
        if( *(ptr++)==' ')
           return ptr;
    return 0;
}
```

The caller could then determine how many characters are before the space like this:

```
n = find_space(test)-test;
```

Because in most circumstances pointers and array names work the same, you can easily split a larger array into a smaller array like this:

```
char all[40] = "Hello World";
char * part2;
part2 = find_space(all)+1;
```

In this case part2[0] is "W" and part2[1] is "o."

Bad Ideas

Pointers can be used to access any memory location by address. For example:

```
int8 * ptr;
ptr = 40;
*ptr = 10;
```

This writes a 10 to memory location 40. You could have also done:

```
*40=10;
```

There is not likely a good reason you would want to do this. Even if it provides some purpose it will likely break when code changes are made, the compiler is updated, or you move to a new chip.

Common Mistakes

When a pointer is declared one must make very sure it is initialized before it is used. The following code will likely do bad things:

```
int * x;
*x = 1;
```

Bounds checking with subscripted arrays are immediately obvious; however, with pointers it is not as easy to see that pointers may go out of bounds. For example, the following code copies one string to another:

```
do {
    *(string2++)=*(string1++);
} while(*(string2-1)!=0);
```

If the string does not have the zero terminator, this code will run on and copy data into other variables until it finds a 0. When dealing with just pointers it is also not as clear if what **string1** points to is large enough for **string2**.

ROM Pointers

The PIC® processors have two separate address spaces, one for normal RAM and special function registers, and a second space for program memory (ROM). Just referring to address 0, for example, is not enough. There is an address 0 in RAM and another one in program memory. This is not true for all processors. A PC, for example, has program memory, RAM, and more all in a single address space. What the PIC® calls special function registers are in their own address space on a PC.

When it comes to defining a pointer, we need to indicate what address space the pointer points to. By default, RAM is assumed. A pointer to ROM looks like this:

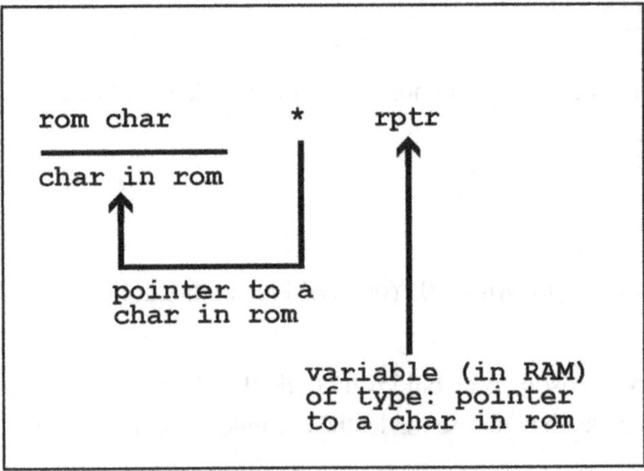

Figure 10.2: Dissection of a point declaration.

```
rom  char  * rptr;
```

Note that the pointer itself (**rptr**) is in RAM. The pointer indication (*) refers to the previous type and the whole thing is the type of what follows (see Figure 10.2).

What if we want the pointer variable also in ROM? Then it looks like this:

```
rom  char   *   rom   rptr;
```

Note that in both of the above, the order of **rom** and **char** does not matter.

Here is an example way a ROM pointer might be used. These strings are better kept in ROM since they don't change.

```
rom char msg_power[]           = {"Enter power Level: "};
rom char msg_minutes[]         = {"Enter time in minutes: "};
rom char msg_safety[]          = {"Unit halted for safety "};
rom char msg_done[]            = {"Operation complete"};
rom char * rom messages[4]     = { msg_power, msg_minutes,
                                   msg_safety, msg_done };
```

This may be done to keep all the test strings in one place so they are easy to change. You may also have cases where this technique makes it easier to deal with multiple languages in the same program. In that case there is an array of pointers for each message and each element of the array is a different language.

User-Defined Memory

We now know the default memory space is RAM and the CCS C compiler as well as others support an address space in ROM. There is an extension to the C language for embedded use that allows the programmer to define a new address space. This space can be anywhere. For example, an LCD display with internal memory, an external serial EEPROM, or even a virtual device at the end of an internet connection.

To define a new memory space you need a name for it (use like **rom** is used) and you must supply two functions, one for reading from the memory and one for writing to the memory. You also give the numeric range of memory addresses for your new space. The **addressmod** is used to define the space. Here is a full example:

```
int8  simulated_memory[100];

void read_simulated( int addr, int * data, int count ) {
    for(int i=addr; i<(addr+count); i++)
        *(data++) = simulated_memory[i];
}

void write_simulated( int addr, int * data, int count ) {
    for(int i=addr; i<(addr+count); i++)
        simulated_memory[i]= *(data++);
}

addressmod( myspace, read_simulated, write_simulated, 0, 99 );

...

myspace int32 x;
x = 0x12345678;

// A 0x78 is written to simulated_memory[0], a 0x56 to simulated_memory[1]
// and so on...
```

Compatibility Note

The **rom** qualifier is not part of the C standard. In fact for many compilers (such as a PC compiler) the program memory and data memory are in the same space. PIC® compilers need some way to differentiate memory spaces. The CCS C compiler uses **rom**.

addressmod is one of the many extensions to the C language in the IEEE Embedded C standard (ISO/IEC TR 18037). Getting agreement on standard additions is a very slow process so this feature is not in many compilers.

Over the Hill

Now is a good time to reread this chapter. A solid understanding of pointers is what separates the amateur from the professional programmer. If there is any area of this chapter that you do not have a solid understanding of, then it will impact your ability to review code and will slow you down in finding frustrating bugs in your code. On the other hand, if you fully grasp these concepts everything else from here is easy. In other words it is downhill from here.

Summary

- Processors access memory by numeric address.
- The `&` address-of operator returns the address in memory of a variable.
- The `*` indirection operator returns the contents from a specified memory address.
- Pointers are memory addresses with a specific type representing the object at the pointer address.
- Addition and subtraction on pointers are multiplied by the size of the object pointed to.
- The `x[y]` operator for arrays is the same as `*(x+y)`.
- The `x->y` operator for structures is the same as `(*x).y`.
- It is possible to declare a pointer to a function and to call a function by the pointer.
- The `rom` qualifier may be used to create a pointer to ROM.
- `addressmod` may be used to create a user-defined address space in addition to the default RAM address space and ROM address space.

Exercise 10-1

Objective: Learn how to use C pointers by coding a linked list. In addition, learn how to use C pointers to directly access memory address.
Requires: E3 module, USB cable, PC.

Steps/Technical Procedure	Notes

1. Given the following data definitions:
`#define UNUSED 0xFFFF`
`typedef struct {`
`int32 id_number;`
`nodeptr next;`
`} node;`
`typedef node * nodeptr;`
`node list[20];`
`int next_node=0;`
`nodeptr first=UNUSED;`
Write a program that asks the user to enter an ID number. Your code will then insert the ID into the next available node in the list identified by next_node. The code will manage a list of nodes starting with first and ending with a node that has its next pointer as UNUSED. Each node has a next pointer to the next node in the list. The list is kept in numeric order so when a new entry is made, the code looks through the list for the right spot and adjusts the pointers so the new node is inserted in the right spot. After insertion the program will display all the IDs in numeric order. The list itself has the entries in the order they were entered.	

Let us now assume that for an ID that looks like 1234567 the 45 digits are encoded such that they represent the person who issued the ID. Add to the data structure a member called next_ issuer and add a variable first_issuer. Now keep two linked lists, both updated on each entry. One for the IDs in order and the other sorted by issuer. Display both lists after each entry.	
2. Write a program that will repeatedly prompt the user to enter R for RAM or P for program memory. Then ask the user for an address (in hex). Finally display to the user the contents in hex of that memory address from PIC® memory. RAM is shown as 8 bits and program memory as 16 bits.	

Quiz

(1) For the following statement, how can you describe the value in x?

```
X = &A + &B;
```

(a) The sum of A and B
(b) The contents of memory at the sum of A and B
(c) A pointer to both A and B
(d) This is not legal C
(e) X is garbage of no use

(2) As an alternative to access a byte array element `x[y]`, which of the following will not work?
(a) `*(&x+y)`
(b) `*(&x[1]+y-1)`
(c) `*(x+y)`
(d) `(int8)*((int8 *)x+(int8)y)`
(e) `x[y]`

(3) What will the following code do?

```
struct { int8 a,b,c,d,e; } s,t,u;
int8 * p;
p = &s;
*(p+2)=5;
```

(a) Writes 5 to u.a
(b) Writes a 5 to s.c
(c) Writes a 5 to s.a
(d) Writes a 5 to an unknown memory location
(e) Syntax error. This is not legal C

(4) What will the following line of code do?

```
x = *&y;
```

(a) Assigns the contents of the memory location in y to x
(b) Assigns y to x
(c) Assigns the memory location of y to x
(d) Assigns the memory location in y to x
(e) Syntax error. This is not legal C

(5) For the following code, what is the value of x?

```
int16 x;
int16 * p;
int16 a[5];
#locate a=0x100
a[0]=3;
a[1]=4;
p=a;
x = p+1;
```

(a) 3
(b) 4
(c) 0x101
(d) 0x102
(e) Syntax error. This is not legal C

(6) What is the value of x in the following code?

```
struct { int8 a,b,c,d,e; } s,*ps;
int8 x;
ps=&s;
s.a=2;
s.b=4;
x = ps->(a+1);
```

(a) 0
(b) 2
(c) 3
(d) 4
(e) Syntax error. This is not legal C

(7) What is the value of x in the following code?

```
struct { int8 a,b,c,d,e; } s, t,
*ps;
int16 x;
s.a=1;
s.b=2;
t.a=3;
t.b=4;
ps=&s;
s.a=ps->b;
x = ps->a;
```

(a) 1
(b) 2
(c) 3
(d) 4
(e) Syntax error. This is not legal C

(8) Function pointers may be used for which of the following situations?
(a) To pass a callback function to another function so the callback function is called during the execution of the function
(b) To have an array of functions whose call order is changed by changing the order of the function pointers in the array
(c) To have a global error handler function pointer where at run time the handler function used can be changed based on the current mode
(d) None of the above
(e) All of the above

(9) For the following declaration, in what memory space is p and what p points to?

```
char  * rom p
```

(a) p is in ROM and it points to a RAM location
(b) p is in ROM and it points to a ROM location
(c) p is in RAM and it points to a ROM location
(d) p is in RAM and it points to a RAM location with a ROM address
(e) Syntax error. This is not legal C

(10) For the following code, what is the value of x?

```
rom int8  x[4] = {1,2,3,4};
rom int8 * rom p = {&x};
int8 a[3] = {9,8,7};
x = a[ p[1] ];
```

(a) 2
(b) 3
(c) 9
(d) 8
(e) 7

CHAPTER 11

Built-In Functions

Traditionally C compilers come with libraries of functions that can be called by the programmer. This is done by including a header file with the function prototypes. After a program is compiled, it is linked with the library supplied with the compiler.

To accommodate direct calls to an operating system, many compilers offer a syntax extension to C that would allow calling functions not part of a supplied library, but rather external to the program.

The CCS C compiler supplies library functions in several unconventional forms in order to best accommodate the PIC® architecture.

- The *#device* directive used to select a chip will cause a number of functions to be available for the specified chip. For example, setting up interrupts or accessing special CPU instructions. Prototypes for these functions are included in the device header files supplied with the compiler. The **`#device`** is also in the device header file, so simply including that header file (like **`16F877.h`**) gives you access to all those functions and the **`#define`** constants used by those functions (like PIN_B0).
- The *#use* library(options) syntax is used for cases where the compiler generates functions on the fly at compile time according to specific needs. For example, the **`#use rs232(uart1,baud=9600)`** directive will create a **`getc()`** function to get a character from the UART (serial transceiver) and a **`putc()`** function to send a character. The generated functions are specific for this chip and baud rate. A generic library that works for all possible combinations would take a lot of memory. This technique allows for generating a function with just what the programmer needs. Prototypes for these functions are also included in the device header files supplied with the compiler.
- ANSI C requires certain functions to be in all compilers. The normal way to access those functions is to include a header file (like **`stdlib.h`** for the most common standard library functions) and then link to the actual library. The CCS C compiler simply inserted in the standard library **`.h`** files the code for the function instead of a prototype requiring a separate link. These standard **`.h`** files are in the drivers directory of the compiler install.
- CCS has, in addition to the standard functions, many more functions in C source code form in other files in the drivers directory that may be **`#include`**d. Some of these are **`.c`** files. All are include files. For example, the **`2416.c`** file has functions for communicating with a 2416 serial EEPROM (electrically erasable programmable read only memory).

Embedded C Programming. http://dx.doi.org/10.1016/B978-0-12-801314-4.00011-9

In summary, a given function is available by either **#include**'ing a file or by using the **#use** directive. In the compiler reference manual for each function it will indicate what you need to do to make the function available.

The remainder of this chapter will cover some of the popular functions grouped by function. This does not replace the reference manual and does not list all functions. There is no point to memorizing these functions, just be familiar with the kinds of functions that are available.

Math

The C standard has a number of math functions available from **math.h**. These functions operate on floats (see Table 11.1).

stdlib.h also has a few math functions that operate on integers, as shown in Table 11.2.

Memory

The C standard has functions that directly operate on memory. These can be dangerous to use so be careful with the ones outlined in Table 11.3.

Some example uses:

```
memset( bigarray, 0, sizeof(bigarray));
if( !memcmp(&oldstruct, &newstruct, sizeof(oldstruct)) )
        cout << "Data changed from old to new" << endl;
```

Table 11.1 Math functions in math.h.

sin(), cos(), tan()	Standard trig functions, all work in radians
asin(), acos(), atan()	Arc trig functions
sinh(), cosh(), tanh()	Hyperbolic trig functions
atan2(x,y)	Finds the arc tangent of x/y
log(), exp()	Natural logarithm and natural exponent
log10()	Base 10 logarithm
sqrt()	Square root
ceil(), floor()	Round a float up and down to an integer
pow(x,y)	Returns X^Y
fabs()	Absolute value
fmod(x,y)	Returns the remainder of x/y
modf(x,&y)	Returns the fractional part of x and changes y to be the integer part (as a float)

Table 11.2 Math functions in stdlib.h.

abs(), labs	Absolute value on an int and a long. The CCS compiler allows abs() on any type
div(x,y), ldiv(x,y)	Return a structure with the division and remainder of x/y for an int and a long

Table 11.3 Memory functions.

memcpy(dest,source,size);	Copies size bytes from source to destination. Both are pointers
memmove(dest,source,size);	Like memcpy; however, it is a smarter version that will make the copy work even if an area in the source and destination overlap
memcmp(ptr1,ptr2,size)	Compares two memory areas and returns true or false
memset(ptr,value,size);	Fills an area size bytes long with value
memchr(ptr,value,size)	Returns a pointer to the first byte at or after pointer that is equal to value

Standard C also has a function that can return the offset in bytes of a member of a structure. It works like this:

```
struct { int8 a,b,c,d; } s;
offset = offsetof( s, c );
```

In this case offset is 2. The CCS C compiler has a similar function for finding a bit offset:

```
offset = offsetofbit( s,c );
```

And in this case offset is 24.

Dynamic Memory

C compilers have a set of functions to deal with *dynamic memory management*. These are frequently used functions in a PC environment, but not so much on a PIC®. So far we have covered memory allocated globally and locally. A third category of memory is allocated at run time, not to specific variables but rather as pointers to blocks of memory. These blocks can be allocated, used, and then when done returned to what is referred to as the *heap*. The heap is a collection of available memory for dynamic use.

Consider a system where buffers are needed for each communication port, each output device, and each user keyboard. The system can have up to, say, 10 of each type of device, but only 12 devices in total. There is not enough memory for 30 buffers but there is for the 12. The problem is at compile time, it is unknown what devices the user will connect. The solution is the buffers are added at run time as the devices are detected. It is an elegant solution to a common problem. The issue for a PIC® comes in because the heap management code can be complex and take a lot of ROM. Each time memory is returned it must be combined with adjacent code, and to allocate memory an algorithm must find the best fit for the requested size to reduce fragmentation. The ROM to handle this can be expensive. Another problem is the PIC® instruction set is not good at dealing with data pointed to by pointers. This too takes a lot of ROM. Finally there is not a good way to handle an out-of-memory situation in a PIC®. On a PC an out-of-memory error simply pops up.

On a PIC® the above problem is usually dealt with by allocating 12 buffers and making each buffer a union of three buffer structures for the three device types.

However, for those who want to do dynamic memory management on a PIC® these functions are available. To save ROM the functions are not in the standard place (**stdlib.h**) but rather in **stdlibm.h**. Only those who want to use those functions should include the **stdlibm.h** file. The functions are in Table 11.4.

A Few More Cool Functions

The C standard also includes in **stdlib.h** the functions listed in Table 11.5:

Table 11.4 Dynamic memory functions.

ptr=malloc(size);	Gets size bytes from the heap and returns a pointer to the memory
ptr=calloc(count, size);	Like malloc, but the size of memory allocated is count*size. Used to allocated a certain number of items of a specified size
ptr=realloc(ptr, size);	Attempts to change the size of a previous block of memory to a new size. Returns 0 if it could not be done
free(ptr);	Returns a block of memory to the heap

Table 11.5 Additional functions in stdlib.h.

srand(n);	Starts a random number generator with n as the seed
value=rand();	Returns a random number from 0 to 255
ip=bsearch(&key, base, num, width, comparefunct)	Returns a pointer to an occurrence of key in the array pointed to by base. Num is the number of array elements and width is the size in bytes of each element. The comparefunct is passed to do the actual comparison. The binary search algorithm is used. This means the elements must be in order for it to work
qsort(base, num, width, compare-funct)	Sorts an array of num elements of width bytes each

Random numbers are very difficult to obtain on a microprocessor. Using **srand()/rand()** you will always get the same numbers if you pass it the same seed. Here are some techniques to get a better seed:

- Save the last value gotten from **rand()** in data EEPROM and use that for a seed on the next power-up.
- If there is any user input (like a button press), then after the press read a timer and use the timer value as the seed.
- If an analog input is connected to something that is always changing, then read the digitized analog value and use that for the seed.

Variable Argument List

C has a mechanism to allow the number of and type of parameters to a function to be variable or up to the caller to decide. Here is the syntax:

```
void funct(...)
```

The ... indicates an undetermined parameter list. You can have both fixed and variable parameters together; however, all the fixed parameters must be at the beginning of the list. The use of this construct is somewhat complicated and especially on the PIC® is not usually worth the trouble. The syntax allows the definitions of functions like **printf()** where the number of and type of arguments are not fixed. The functions used to get the parameters are in Table 11.6.

A full example:

```
int32 sum_up(...) {
```

Table 11.6 Variable argument list functions.

nargs()	Gets the count of arguments
va_start(ptr,count)	Used to initialize a pointer to the arguments. Count is the max number of arguments
va_arg(ptr,type)	Used to get the next argument. The expected type of the argument must be passed in so the function knows how many bytes to get
va_end(ptr)	Used to end the process of getting arguments

```
int32  sum_up( ... ) {
   int32 sum = 0;
   int i;
   va_list argptr;   // special argument pointer
   va_start(argptr,nargs());   // initialize argptr
   for(i=0; i<nargs(); i++)
      sum = sum + va_arg(argptr, int);
   va_end(argptr);   // end variable processing
   return sum;
}

void main(void) {
   int a,b,c,d;
   a=1;
   b=2;
   c=3;
   d = sum_up( a,b,c );
   count << d;    // outputs 6
}
```

Text Input/Output

The C standard has a number of functions to deal with character input and output. These functions are intended for both a keyboard/monitor type of I/O and for disk file I/O. For a PIC®, the CCS C compiler uses these functions primarily for RS-232-like I/O or to other user-defined I/O. Most functions have two versions, one where you specify a stream name (like **fprintf()**) and another where the default stream is used (like **printf()**). If using an RS-232 port on the PIC®, these functions are made available using a directive like this:

```
#use rs232( baud=9600, xmit=PIN_C6, rcv=PIN_C7, stream=PORT1 )
```

A character could be sent out of the port using a call like this:

```
fputc( 'A', PORT1 );
```

Most C compilers have a **stdio.h** file that includes these functions and functions to open disk files. In the PIC® the **sdtio.h** is usually not needed since the functions appear with **#use RS232**.

The basic text I/O functions are in Table 11.7.

These are fairly straightforward functions except for the more complex **printf()**. The basic form looks like this:

```
printf( format-string, argument1, argument2... );
```

Table 11.7 Text I/O functions.

putc(c) fputc(c, stream)	Send a single character out
putchar()	Same as putc()
c=getc() c=getc(stream)	Receive a single character in
c=getchar() c=getch()	Same as getc()
puts(s) fputs(s,stream)	Output a string followed by a \r\n
gets(s) fgets(s,stream)	Input a string terminated by a \r
printf() fprint()	Formatted output
kbhit() kbhit(stream)	Return true if a character has been sent but not yet received
aeesrt(cond)	Must to include assert.h to use this function. It outputs a error message if the condition is true. For example: assert(count>100);

Table 11.8 printf() format specifiers.

%u	unsigned int
%lu	unsigned long
%d	signed int
%ld	signed long
%f	float
%e	float output in exponential form
%c	char
%s	string

In the format string are format specifiers (for example `%u`) that refer to each argument in order. For example:

```
printf( "Position is x=%u, y=%u", x, y);
```

In this case the first `%u` is replaced with the string representation of `x` and the second `%u` is replaced with the string representation of `y`. The other characters of the string are output as is. To get a `%` in the string use `%%`. The most commonly used format specifiers are shown in Table 11.8.

Format specifiers can include a width. This is how many characters to output. For example:

%3u would output 12 as " 12" (one leading space to make 3 chars)

A zero after the `%` will cause the leading spaces to be leading zeroes:

%03u would output 12 as "012"

A minus after the `%` causes left-hand justification, like this:

%-3u would output 12 as "12 "

For floating-point numbers you can specify a total width and the number of digits to the right of the decimal point like this:

Table 11.9 Output manipulator list.

hex	Hex format numbers
dec	Decimal format numbers (default)
setprecision(x)	Sets number of places after the decimal point
setw(x)	Sets total number of characters output for numbers
boolalpha	Outputs int1 as true and false
noboolalpha	Outputs int1 as 1 and 0 (default)
fixed	Floats in decimal format (default)
scientific	Floats use E notation
iosdefault	All manipulators to default settings
ensl	Outputs CR/LF
ends	Outputs a null ('\000')

%7.2f would output 123.456 as " 123.45"

The newer way to do formatted output is using the stream operators with **cin** and **cout**. To use this capability **ios.h** must be included. Output looks like this:

```
cout << data;
```

cout may be any stream name from a **#use RS232**. Using **cout** itself directs the output to the default stream. It may also be a string name or the name of a function to receive the data. In the case where a function name is used, the function is called with each character. Multiple outputs may be done in the same statement like this:

```
cout << data1 << data2 << data3;
```

In addition to constants and variables, manipulators may be used where data appears. Manipulators do not output any data but affect how the future outputs on the line are treated. For example, to output a number in hex the hex manipulator is used like this:

```
cout << hex << data;
```

The valid output manipulators are shown in Table 11.9.

Input is similar, with the basic statement looking like:

```
cin >> data;
```

For input, the processor will wait for a return to be entered before moving to the next statement. By default the data input is echoed out. If there are multiple inputs in the same statement then the traditional method is to separate the items input with a space. For example:

```
cin >> data1 >> data2 >> data3;
```

The manipulators for input are shown in Table 11.10.

In addition to any stream name, **cout/cin** can be a string or function name. When a string is used then the data is written to or read from the string variable. For a function, when output

Table 11.10 Input manipulator list.

hex	Hex format number
dec	Decimal format number (default)
noecho	Suppresses echoing
strspace	Allows spaces to be input into string
nostrspace	Spaces terminate string entry (default)
iosdefault	All manipulators to default setting

is done, the function must accept a single character and it will be called for each character to be output. For input the function must return a char. The functions may be used for custom I/O devices like an LCD or keypad. Finally, it is possible to do input and output in the same directive to cause incoming data to be echoed as it is input. Here are some examples:

```
cout << "Value is" << hex << data << endl;

cout << "Price is $" << setw(4) << setprecision(2) << cost << endl;

lcdputc << '\f' << setw(3) << count << "  " << min << "  " << max;

string1 << setprecision(1) << sum / count;

string2 << x << ',' << y;

cout << "Enter title: ";
cin >> strspace >> title;

cin >> data[i].recordid >> data[i].xpos >> data[i].ypos >>
       data[i].sample ;

string1 >> data;

lcdputc << "\fEnter count";
lcdputc << keypadgetc >> count;     // read from keypad, echo to lcd
                                     // This syntax only works with
                                    // user defined functions.
```

At this point you are wondering about the syntax used for the stream operators. It would appear to violate some of the basic C rules learned thus far. This capability is from C++ and in C++ you can define your own data classes and tie specific C functions to the normal C operators. For example, you could redefine how C does an add with the **+** operator when your specially defined data is involved. This is a kind of neat capability, but not practical to implement on a PIC® type of processor. C++ has **cout** and **cin** defined as a special data type and to use strings they make you use a special string definition as well. This is how they make the magic work. For the CCS C compiler, the **>>** and **<<** operators simply look for the specific syntax described above

Table 11.11 Sampling of standard C constants.

limits.h	INT_MIN	Smallest integer for a "int"
	INT_MAX	Largest integer for an "int"
	Similar constants are defined for all the standard types, signed and unsigned	
float.h	FLT_MIN	Smallest floating-point number
	FLT_MAX	Largest floating-point number
	FLT_MIN_10_EXP	Smallest exponent allowed for a float (in base 10)
	FLT_MAX_10_EXP	Largest exponent allowed for a float (in base 10)
	Many similar constants are defined for both float and double	
stddef.h	size_t	The data type that sizeof() returns
	ptrdiff_t	The data type that results when two pointers are subtracted
	NULL	The string terminator character (\000)
	offsetof()	Macro to find the offset of a member in a structure

and emulate the C++ operation. Essentially, instead of an elegant expression implementation the compiler has extended the definition of a statement to cover these special cases.

Implementation Constants

The C standard mandates certain include files be provided with the compiler that have **#define**s in them to identify implementation-dependent constants. Table 11.11 gives a partial list to help understand what can be found in these files.

Compatibility Notes

Most compilers require the inclusion of stdio.h to get the above functions, like **putc()** and **getc()**. The CCS C compiler uses the **#USE RS232** to make those functions available. Most PIC® compilers will have some quirky way to use functions like this.

The **ios.h** functions are not part of standard C compilers. They come from C++ and are available in only some compilers.

The remaining built-in functions in this chapter and most of the built-in functions that will be introduced in the remaining chapters are unique to the CCS C compiler because they deal with PIC®-specific capability and/or functionality.

Bit and Byte Manipulation

The PIC® has instructions that allow for efficient bit and byte manipulation. The CCS C compiler has a number of functions to take full advantage of those capabilities by making C level functions available. Table 11.12 is a summary of those functions.

Table 11.12 PIC®-specific built-in functions.

bit_set(v,b);	Sets bit number b in variable v to a 1
bit_clear(v,b);	Sets bit number b in variable v to a 0
x=bit_test(v,b);	Returns the value of bit b in v
x= make8(v,n);	Returns byte number n from variable v. S. A 0x12345678 then make8(v,3) is 0x12. The low byte number is 0
x= make16(h,l);	Returns an int16 from two int8 variable v. If V was make 16(0x11, 0x22), will return 0x1122
x=make32(a,b,c,d);	Returns an int32 from four int8 variables. Can also be called with any combination of int8 and int16 variables that add up to 32 bits
rotate_left(ptr, n);	Rotates n bytes by one bit to the left. The low-order byte is the first one pointed to. For example, if ptr is an array with 3 bytes (0x03, 0x82, 0x81) and n is 3 will view the data as 0x818203 and the rotate of bits will result in 0x030407, so the array will have: 0x07, 0x04, 0x03
rotate_right()	Same as rotate_left(), just the other direction
bout=shift_left(ptr, n, bin);	Line the << operator except you have control over the bit that gets shifted in (<< uses 0) and you can get the bit shifted out. For example, if x was 0x82 and you did shift_left(&x,1,1) you would get in return a 1 and the data in x would be 5
bout=shift_right(ptr, n, bin);	Same as shift_left(), just the other direction
swap(bvar);	Swaps the upper and lower nibbles of a byte. For example, if bvar was 0x12 it would change bvar to 0x21

Non-volatile Memory

Non-volatile memory is memory that retains its values even after power is removed. The classic example is EEPROM (electrically erasable programmable read only memory). Many PIC® chips have a small area of EEPROM available for storage of data that needs to be retained. An example would be the last channel and volume setting on a TV. Even after power is lost and restored, you want to be able to remember those settings.

The basic functions in the compiler for this are **read_eeprom()** and **write_eeprom()**. The addresses used start at 0 and go up to the last EEPROM location on the device. An erased chip will have 0xFF in the data EEPROM. For example, the following code might appear at the start of main:

```
volume = read_eeprom( 0 );
if(volume==0xFF){
   volume= 10;
   write_eeprom(0, volume);
}
```

And if the user hits the volume-up button:

```
volume++;
write_eeprom(0,volume);
```

This uses a simple trick of looking for 0xFF to figure out if the EEPROM has ever been used. Another method used would be to sum up all the used EEPROM locations and save that sum in another EEPROM location. This sum would be checked on power-up and if wrong the whole EEPROM is initialized with the factory default values. This method solves a problem where in the middle of writing a multi-byte EEPROM change you lose power and have an incomplete change to the EEPROM.

Another concern you must have when dealing with EEPROM data is there is a limit to the number of writes that can be done over the life of the EEPROM. This might be as low as 100,000 writes. For example, if your program writes to the EEPROM every minute (some kind of counter, or logger) then the chip will stop working after a couple of months. Some EEPROMs have a limit of writes per location; with others the limit is total writes per chip. Most PIC® parts are the latter. Some EEPROMs work better if they are refreshed periodically. For example, `write_eeprom(0, read_eeprom(0));`. If this is done, say, once a year there is a better chance the data will be retained. Most PIC® devices have this recommendation.

One final consideration for EEPROM use is time. Reads are fast but writes can be very slow. It will also take longer to write as the chip ages. Count on about 10 ms for each byte write. New chips will be much faster, but use 10 ms when budgeting time.

For applications that need more EEPROM than the PIC® provides, an external serial EEPROM can be used. It is easy to connect these up with two wires to the PIC®, and the CCS library has drivers ready to use for most serial EEPROM devices. These parts range from 128-byte parts to 262,144-byte parts. Here is an example that uses a CCS C driver for a 256-byte part. It sets all locations to 0 if the first location is FF:

```
#define EEPROM_SDA  PIN_C4   // Sets the pins to use for the driver
#define EEPROM_SCL  PIN_C3
#include <2402.h>   // include driver functions
void main(void) {
   int i;
   if (read_ext_eeprom(0) == 0xFF)
      for(int i=0; i<=255; i++)
         write_ext_eeprom(i, 0);
}
```

Many (but not all) PIC® chips also allow a program to write to program memory, ROM. This can be used somewhat like the data EEPROM for programs that need more storage than the data EEPROM provides (256–1024 bytes is typical).

The program memory has some special considerations. First you must make sure the compiler is not going to use the area of program memory you plan to use. The following directive will prevent the compiler from using locations 0x1000 to 0x17FF:

```
#org 0x1000,0x17FF {}
```

The **org** is normally used to force a function at a specific address and this form of the directive prevents anything from going there. Another consideration is each chip has a specific erase block size. The erase operation changes all locations to 1s and the write operation will change 1s to 0s. The following function will write 256 bytes from RAM to location 0x1000 in program memory:

```
write_program_memory(0x1000, ptr2ram, 256);
```

The compiler function will erase a block when you write to the first byte in the block. In this case if the erase block size was 1024 then in addition to the 256 bytes being written you would get 768 bytes of 0xFF erased afterward. If the first address was 0x1001 then the compiler would not do the erase and this would cause problems if the block were not erased. You can manually erase a block like this:

```
erase_program_memory(0x1000);
```

Reading program memory to RAM can be done like this:

```
read_program_memory(0x1000, ptr2ram, 256);
```

You can also get serial Flash parts (like the serial EEPROMs) that allow for many megabytes of external storage. Like the program memory above, these parts have an erase block size and the size seems to get bigger as the total memory gets larger. Again, the CCS compiler has predefined drivers for many of these parts.

The following code uses the **read_program_memory()** function to calculate the checksum of the entire program memory. Some programs will do this periodically to ensure there was no corruption of memory.

```
int16 checksum=0;
int16 value;
for(i=0; i<getenv("Program_memory")-1; i++) {
   read_program_memory(i, &value, sizeof(value))
   checksum+=value;
}
```

The **#rom** and **#id** directives described in Chapter 3 are able to save a compile-time-calculated checksum in memory.

Watchdog Timer

All the PIC® parts have a feature referred to as the watchdog timer (WDT). This timer runs on its own clock (not very accurate) and after a certain period of time will reset the processor. To prevent the processor from being reset the program must reset the timer before it expires. For example, if the timer is set up for 2 seconds then the program would want to restart the timer at most every 2 seconds, but to be safe would probably do it every second. If the program gets stuck somewhere then the watchdog timer will expire and will reset the chip, causing the program to start over in main.

Usually this is a fail-safe trigger that will account for errors in the software that cause it to get stuck. On the assumption that the code mostly works but there is a rare case where it gets stuck, this will reset the chip and allow the device to function to some degree.

Sometimes programmers will count on the WDT to reduce program logic. For example, consider a unit needs to communicate with several external devices to operate and one of those devices is unplugged—the unit can't do anything. One way to handle this is to put in code to always check for a response within a certain period of time and if the response is not received then try to restart the device. Another way is to allow the code to hang waiting for the response and let the WDT restart the chip. Here is an example program:

```
void main(void)    {
   if(restart_cause()==WDT_TIMEOUT)
     cout << "Restarted processor due to watchdog timeout!" << endl;

   init_hardware();
   setup_wdt(WDT_2304MS);

   while(TRUE) {
      restart_wdt();
      gather_data();
      process_data();
      check_for_problems();
   }
}
```

This program assumes each loop through the main while will take less than 2 seconds if all is running well. Sometimes the **restart_wdt()** calls are spread throughout the code where needed based on processing time. It is not a good idea to restart the WDT inside a timer interrupt. That prevents a hang in the main program from triggering a timeout.

To save power some programmers, when they don't have anything more to do at the moment, will put the chip to sleep and then the WDT wakes the chip up, say, in a tenth of a second.

The program might then do 10 ms' worth of work and go back to sleep. The power draw during sleep is very low so this can save a lot of power for, say, a battery-operated device.

Delays

You have already worked with the **delay_ms()** function. That function came from a directive in the **e3.h** file that looks something like this:

```
#use delay(crystal=10mhz)
```

This directive tells the compiler that the chip is using a 10-MHz crystal as the clock and the compiler will generate the following delay functions:

delay_ms(n);	Delays n milliseconds
delay_us(n);	Delays n microseconds
delay_cycles(n);	Delays n instruction clocks (¼ crystal speed on most devices)

The **#use delay** also will set configuration bits and initialize the PIC® oscillator for the requested speed. Some PIC® parts have an internal PLL to allow the chip to run faster than the crystal. In this case the directive looks like this:

```
#use delay(crystal=10mhz, clock=40mhz)
```

Some PIC® chips also have an internal clock that can be set to a number of frequencies. To use an internal clock, the directive looks like this:

```
#use delay(internal=8mhz)
```

Multiple Clock Speeds

Sometimes a program will need to run at different speeds. This can become very complex with compilers, built-in functions that depend on knowing the clock speed. The reason for dual speeds is usually because a high speed is needed to do a lot of processing or to communicate with another device at some high rate. The slow speed is used for idle times to conserve power. The faster the clock, the more power is required. The general format looks like this:

```
#use delay(crystal=31khz)          // slow speed functions here
#use delay(crystal=20mhz)          // all high speed functions here
```

With this structure the built in functions called will assume the right clock speed; however, the speed will not actually change. To change the speed a call must be made to **setup_oscillator()**. It looks like this:

```
setup_oscillator(OSC_31KHZ);
```

The parameters vary a lot between chips so be sure to check the device header file. Make sure to put library generators like **#use rs232** under the right **#use** delay.

A Few More Standard Functions

The **getenv()** function is a standard C function that is used to grab parameters set outside the program. These are traditionally operating system variables that could be set up before a program runs and then the program can use that data when it runs. Typical uses would include setting a specific directory path or device name (like a printer). None of this makes sense on a PIC®. The CCS C compiler instead uses this function to obtain information about the PIC® processor being used or compiler settings that have been set.

For example, use **getenv("PROGRAM_MEMORY")** to find out how much program memory the PIC® has. Above we discussed how PIC® devices have a specific erase block size. Find the erase block size by using **getenv("FLASH_ERASE_SIZE")**. You can even find the address for a PIC® register with something like **getenv("SFR:STATUS")**. **getenv()** can be used in preprocessor directives as well as in C code. It is treated like a preprocessor macro. There are many options in **getenv()**; check the reference manual for all of them.

You can restart the chip with **reset_cpu()** and you can force the chip to go into a sleep mode with **sleep()**. The chip can wake from a sleep for some interrupts if they were enabled or by a watchdog timeout. When you wake from **sleep()** with an interrupt the program execution continues normally. When you wake from a watchdog timeout the processor resets.

At the start of **main()** you can call **restart_cause()** to find out why the chip was reset. The constants for a specific chip are in that chip's, header file. For example, to find out if the chip was reset due to a watchdog timeout do this:

```
if(restart_cause()==WDT_TIMEOUT)
   cout << "Watchdog timer reset";
```

Coming Up

The string functions will be covered in detail in Chapter 12. There are more built-in functions primarily dealing directly with PIC® hardware features. Those will be covered in future chapters in detail. The groups of function being deferred are:

Timers and counters
Discrete I/O on PIC® pins
Pulse measuring and generation
Interrupt functions
Analog voltage functions
I²C bus functions
SPI bus functions.

Summary

- Making use of built-in functions that come with a compiler can save a lot of time and make it easier to migrate code to a new PIC® chip.
- Some built-in functions are associated with the chip selection (**`#device`**), some are generated as part of a dynamic library (**`#use`**), and some are supplied as source code in include files.
- ANSI C defines a number of required math, memory, and special functions that all compilers have.
- It is possible to have a variable number of and types of arguments to a function when special built-in functions are used to obtain the parameters.
- The standard C text functions can be used on a PIC® by directing those functions to a serial port or to any kind of interface (like LCD and keypad) the programmer defines.
- PIC® bit and byte manipulation functions extend the standard C methods of dealing with bits and bytes.
- Most PIC® parts have a small area of non-volatile memory that can be used for data storage across power cycles.
- All PIC® parts have a watchdog timer that can be used to reset the processor after a period of time has gone by without the code resetting the timer.
- The **`getenv()`** function is a powerful tool to obtain information about the processor the code is being compiled for.
- The compiler reference manual and help file should be used to get details on all the built-in functions.

Exercise 11-1

Objective: Become familiar with the use of compiler built-in functions. The reader will also learn to use the IDE graphing capability and introduce parity checking and error collection.
Requires: E3 module, USB cable, PC.

Steps/Technical Procedure	Notes
1. The serial input/output monitor program you have been using to interact with the PIC® programs has a feature that allows you to display graphs. On the PIC® side functions to help do the graphing are in an include file named **`graph_siow.h`**. Before this file is included you can **`#define`** a number of parameters such as the title for the graph and the ranges of the X and Y axes. Help for using the functions is at the top of the include file. Write a program that uses **`init_graph()`** and **`graph_point()`** to display a sine wave for 0 degrees to 719 degrees. Remember the functions in **`math.h`** are in radians so you will need to do a conversion. The range for the Y axis should be −1.0 to 1.0.	

(continued)

Steps/Technical Procedure	Notes
2. Now modify the program to graph two lines, one sine and the other cosine.	
3. One method of sending data over unreliable communication channels is to add parity bits to the data. For example, you could take an 8-bit byte and if there are an odd number of 1s in the byte the parity bit would be 1 for the byte. To perform error correction in addition to error detection you calculate a column parity by counting the 1 bits in the same position of each byte. Consider 8 bytes in binary stacked one on top of another. You have an 8 by 8 matrix of 1s and 0s. Each row and column has a parity bit, for a total of 16 parity bits. This means there are an extra 2 bytes send with every 8. In this case if there is a 1-bit error one of the column parity bits will be wrong and one of the row parity bits will be wrong. You now know which bit is an error and since each bit location can only be one of two values and you know it is wrong, it can be fixed. If only one parity bit is bad then the error may be the parity bit itself. If more than one is bad in the row list or column list then there is a multi-bit error and it cannot be corrected. Write a program to accept 8 bytes of data (in hex) and 2 parity bytes. The first parity byte has the row parities with the low bit representing the first byte (or row). The second parity byte is the column parities. After entry of the data, tell the user what the corrected 8 bytes are or display an error indicating the data cannot be corrected.	
The following is sample data: (see table below) Sent: 41 42 43 44 45 46 47 48 34 08 Example 1-bit error that should get fixed: 41 43 43 44 45 46 47 48 34 08	

Data	Row Parity
01000001	0
01000010	0
01000011	1
01000100	0
01000101	1
01000110	1
01000111	0
01001000	0
00001000	col parity

Quiz

(1) Before using the **cos()** function what should first appear?
   (a) **float cos(float rads);**
   (b) **#include <math.h>**
   (c) **#include "math.h"**
   (d) **#use math**
   (e) Nothing, it is a built-in function

(2) What will happen with the following code?

```
int32  i=90;
float f;
f = sin(i);
```

   (a) Error because **i** is not a float
   (b) Error because **i** must be in radians
   (c) Error because you cannot assign an integer to a float
   (d) Error because functions must appear in expressions
   (e) No compilation errors here

(3) What is wrong with the following code?

```
struct  {  int8 a[3];
           int32 b;
           int8 c; }  s1, s2;
      memcpy(s1, s2,sizeof(s1.a)+sizeof(s1.b)+sizeof(s1.c))
```

   (a) No data has been written to **s2** yet
   (b) The size of each member of a **struct** is not always equal to the size of the **struct**
   (c) An **&** is needed before the first two arguments
   (d) All of the above
   (e) There are no problems with this code

(4) **memset()** is often used to initialize an entire array to a single value. Special considerations need to be taken for multi-byte elements of an array. What value can a signed int16 array not be initialized to with **memset()**?
   (a) −1
   (b) 0
   (c) 1
   (d) All of the above are valid
   (e) None of the above are valid

(5) Which of the following reasons to maximize the use of built-in functions is not true?
   (a) It is less work for the programmer
   (b) It makes it easier to use the same code on multiple PIC® parts
   (c) It makes it easier to port code to new PIC® processors
   (d) It makes it easier to port code to other compilers
   (e) It makes the code easier to read

(6) If a developer wants to use compiler built-in functions but he knows his PIC® code will be migrated to another processor and compiler, which of the following is the best way to deal with it at coding time?
   (a) Add a special comment to the end of each line that uses a built-in function unique to the compiler, so this can be searched for in the future
   (b) Encapsulate all unique compiler functions inside functions of your own all in one include file so only that file needs to be changed
   (c) Don't use any compiler built-in functions
   (d) Keep a document listing all locations the built-in functions are used
   (e) Do nothing special, leave it for whoever has to do the porting

(7) On a PIC® before you can use **`putc()`** and **`getc()`**, you need what?
   (a) A stream defined via **`#USE RS232`**
   (b) Include the **`stdio.h`** file
   (c) Include the device header file
   (d) A connection to a PC
   (e) Nothing, they are built-in

(8) Which of the following is probably not a good use for the watchdog timer?
   (a) If a user does not type in an answer to a question within a certain period of time, a warning pops up to tell the user to wakeup
   (b) Want to restart if any device does not respond within a reasonable time
   (c) Want to restart if there is an error in the code that causes the code to be stuck in a loop
   (d) Only need to run for a short period of time and then sleep, letting the watchdog timer wake up the processor to run again
   (e) All of the above are excellent uses of the watchdog timer

(9) If power fails while the following code is executed, which of the following will certainly not be in the data EEPROM?

```
int32 x=0x12345678;
write_eeprom(0,make8(x,0));
write_eeprom(1,make8(x,1));
write_eeprom(2,make8(x,2));
write_eeprom(3,make8(x,3));
```

(a) 0x12,0x34,0x56,0x78
(b) 0x78,0x56,0x34,0x12
(c) 0x12,0x34,0x56,0xFF
(d) 0x12,0x34,0xFF,0xFF
(e) Any of the above could be in EEPROM

(10) What is the one address the **write_program_memory()** should never, ever write to?
(a) The reset vector (zero on most PIC® parts)
(b) The location of **main()**
(c) Any program memory location that is not used
(d) The location where the **write_program_memory()** function is located
(e) Any location can be written to

CHAPTER 12

Strings

A *string* is an array of characters with a zero terminator. C does not have a formal string data type. A lot has already been covered about string constants and string variables. This chapter will cover dealing with strings in your program, including input, output, and manipulation of strings.

A picture of a string in memory is shown in Figure 12.1.

Remember that a string is a pointer and that string==&string[0].

String Copy and Length

The length of the string in Figure 12.1 is 11. It requires a minimum of 12 bytes of memory to include the terminator. The function **strlen()** may be used to find the string length. Before it and most string functions can be used, you must include the **strings.h** file. Figure 12.2 is a whole program that will output 11.

The **strlen()** depends on the terminator in the string. If it is not there, that function could hang or do bad things. The argument to **strlen()** is always a pointer. For example, you could do the following:

```
n=strlen(&string[6]);
```

In this case n would be 5. It is common to play with pointers in this way with strings.

To copy one string to another you can do the following:

```
char string1[15] = "Hello World";
char string2[15] = "ABCDEFGHIJKLMN";
strcpy(string2,string1);
```

The array **string2** will now have in it: H e l l o W o r l d \000 M N

The string is "Hello World" and the **strlen(string2)** is 11. The M and N, although there, don't count for anything.

Embedded C Programming. http://dx.doi.org/10.1016/B978-0-12-801314-4.00012-0

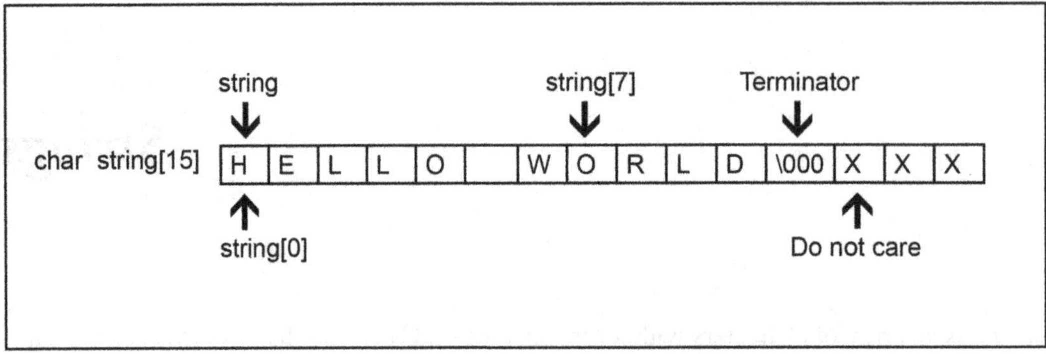

Figure 12.1: Illustration of a string in memory.

```
#include <E3.h>
#include <string.h>

void main(void) {
    char string[15] = "Hello World";
    int  n;

    n=strlen(string);
    cout << n << endl;
}
```

Figure 12.2: Example code for using string.h.

You must be very careful when doing a string copy to make sure you know the source string has a terminator and that the destination string is large enough for the string. No checking is done in the **strcpy()** function to keep things safe. C has another function that is used to make a safer copy. The **strncpy()** function has a third parameter that specifies the maximum number of characters to copy (including the terminator). If the **strncpy()** ends before the source is fully copied then the destination will not have a terminator. Sometimes this is what you want and sometimes it is bad. Be aware of how it works and what you need to do. Here is an example:

```
strncpy(string2, string1, 6);
```

This gives us a "Hello GHIJKLMN" in string2. If **string2** were not previously initialized with the terminator this would be an invalid string. Here is another example:

```
strncpy(&string2[3], &string1[6], 5);
```

to give us the string2 value of "ABCWorldIJKLMN". And one more to include the terminator:

```
strncpy(&string2[3], &string1[6], 6);
```

will give us "ABCWorld", and you would get the same result with any value 6 and up for the third parameter.

String Search

The **strstr()** function is used to search for one string inside another string. The result from **strstr()** is a pointer to the start of the second string in the first string. If the string is not found it returns 0. Be aware that with some PIC® compilers you cannot pass constant strings to these functions as we are about to do, because of the dual address space. Here is an example:

```
char string[15] = "Hello World";
signed int  pos;
pos = strstr(string,"Wor")-string;
```

pos will be 6 after the call. If **pos** were negative it means the substring was not found. Here is code to count all occurrences of "the" in a string:

```
char string[] = "The result from strstr() is a pointer to the"
                "start of the first string in the second string";
int * ptr,
int count;
count=0;
ptr = string;
do {
   ptr = strstr(ptr,"the")+1;
   if( ptr!=1 )
      count++;
} while(ptr!=1);
```

For this example the count will be 3 because the first "The" does not match due to a case difference.

The **strchr()** works like **strstr()** except the second parameter is a character and the search is for the first character in the string. **strrchr()** does the same thing except it searches for the last character match in the string.

The last function in this group is the **strpbrk()** function. It searches the first string for any character in the second string. Here is an example:

```
char string[15] = "Hello World";
signed int  pos;
pos = strpbrk(string,"aeiouyAEIOUY")-string;
```

pos will be 1, matching the **"e"** in hello.

String Compare

These functions perform a comparison of one string to another. The primary function is **strcmp()** and it has two string arguments. The return value is −1 if the first string is alphabetically below the second string. It returns 0 if the strings are identical and 1 if the second string is above the first string. Here is an example:

```
char key1[] = "red";
char key2[] = "green";
cout << strcmp(key1,key2) << endl;
```

outputs a 1 because "red" comes after "green" alphabetically.

The **strncmp()** function is similar except it has a third argument with the maximum number of characters to compare.

The **stricmp()** function is similar to **strcmp()** except it ignores case differences.

The **strspn()** function does a compare as well; however, it returns a count of the number of characters that match, up to the first mismatch. 0 means the first character in each string is different. If the return value is **strlen()** of both strings then they are the same.

strcspn() is like **strspn()** except that it counts the number of characters that do not match, up to the first character that matches.

String Manipulation

The first function we will look at is the **strcat()** (concatenation) function. This function appends the second string to the end of the first string. Simply put:

```
char string1[20] = "Hello";
char string2[] = "World";
strcat(string1," ");
strcat(string1,string2);
```

will give us "Hello World" in **string1**. It should be noted the **strcat()** function returns a pointer to the first string, so one could be cute and do this:

```
strcat(strcat(string1,· ·),string2);
```

Like with **strcpy()**, make sure the destination string (first argument) is big enough to hold all the characters.

The **strncat()** function has a third argument that has a count of the maximum number of characters that will get appended.

The **strlwr()** function has one string argument and the function will convert all uppercase letters in the string to lowercase letters.

String Input and Output

The traditional string input function is **gets()**. Its use is shown below:

```
char line[20];
gets(line);
```

This will enter characters until a return (**\r**) is encountered. A 0 is appended to the string instead of the **\r**. Note that **gets()** does not echo the characters back to the source. This is good if you grab data from, say, a GPS but awkward if a user is typing data. A second warning is **gets()** does no check to see if the string is large enough for the incoming data. Data keeps going into line or whatever variables follow until a **\r** is seen.

The traditional string output function is **puts()**. Used like this:

```
puts(line);
```

This sends out the string (no 0 terminator is sent) followed by a **\r** and **\n**. Some compilers may not send both line terminators. Another popular string transmit function is the formatted print, for example:

```
printf(·The string is:%s \r\n·, line);
```

The stream operators in the CCS C compiler can also be used for easy string input and output. In this case, by default, the input is echoed. For example:

```
char line[20];
cout << "Enter a string: ";
cin >> line;
cout << "The string entered is: " << line << endl;
```

There is a quirk with **cin** and strings, however, in that the entry is terminated when a space is seen, by default. **cin** likes to terminate all variable inputs on a space. To override this you can use the following:

```
cin >> strspace >> line;
```

String Conversion to/from Numbers

The traditional way to convert from a string to an integer is to use the ASCII to integer function, **atoi()**. To use this you must include the **stdlib.h** file. There is also an **atol()** for a long data type, **atoi32** for a int32 data type, and **atof()** for a float type. They are used like this:

```
char  line[10];
int value;
printf("Enter a number: ");
gets(line);
value = atoi(line);
```

To be clear, this changes something like "123" (4 bytes with terminator) to a single-byte 123 (or 0x7B). The stream operator also works on strings instead of the serial input. Using streams, the same code looks like this:

```
char  line[10];
int value;
cout << "Enter a number: ";
cin >> line;
line >> value;          //string to int conversion with stream operator
```

In this case there is no concern about calling the right function for the right data type.

To go the other way (integer to string) there are some traditional functions in **stdlib.h**; however, most people would use a function called **sprintf()** that works like **printf()** except that it writes to a string. Here is an example:

```
char  line[10];
int value;
value=123;
sprintf( line, "%u", value);
printf("The string is: %s \r\n", line);
```

This has the flexibility to create a string using the powerful formatting capabilities of **printf()**. Here is the stream example:

```
char  line[10];
int value;
value=123;
line << value;
printf("The string is: %s \r\n", line);
```

Character Manipulation

C has two functions to convert a character to upper- or lowercase. If the character is not a letter then it is unchanged.

```
c = toupper(c);
c = tolower(c);
```

There are also a number of functions that return TRUE or FALSE depending on what the character is. Those functions are listed in Table 12.1.

Unicode

For a long time a character was 8 bits. There were some odd implementations that used 5- or 6-bit character systems, but as the byte became popular so did the 8-bit character. The problem came as more information was exchanged in foreign languages. It all seemed OK as long as no one had more than 255 characters in their alphabet; however, mixing languages in the same document or application became difficult to handle. The Unicode concept allows for strings of characters in any language. There are different encoding methods. The most popular is called UTF-8. It correlates most closely to the strings in this chapter. The other methods are called "wide," where each character is 4 bytes and the character set is encoded into each character, and there is a multi-byte format where the number of bytes varies for each character. In all cases a char is no longer 1 byte. C provides some functions for Unicode conversions; however there is not a strong standard for this. It is still an evolving concept. You need to know the issue exists but there is no need to cover it in any more detail.

Table 12.1 Standard C character functions.

isalnum(x)	TRUE if x is '0'..'9', 'A'..'Z', or 'a'..'z'
isalpha(x)	TRUE if x is 'A'..'Z', or 'a'..'z'
isdigit(x)	TRUE if x is '0'..'9'
islower(x)	TRUE if x is 'a'..'z'
isupper(x)	TRUE if x is 'A'..'Z'
isspace(x)	TRUE if x is a space
isxdigit(x)	TRUE if x is '0'..'9', 'A'..'F', or 'a'..'f'
iscntrl(x)	TRUE if x is less than a space
isgraph(x)	TRUE if x is greater than a space

Constant String Management

Some programs use a lot of constant strings. For example, any product with some kind of menu system may have dozens of constant strings. It can be helpful to group all the strings together like this:

```
const char MSG_PUMP_IDLE[]    = "Pump is on standby";
const char MSG_PUMP_RUNNING[]= "Pump is running";
const char MSG_PUMP_DOWN[]= "Error detected in pump";
const char MSG_PUMP_OFF[]= "Pump switch is in the OFF position";
```

Grouping messages together in one place, even in a single include file, will make it easier for non-programmers to review the messages and easier for you to change them.

Using an array of pointers to the messages makes the code a little more complex; however, it has the advantage of making it easy to support multiple languages or to use slang for different products or branding. The CCS C compiler does not allow pointers to a **const**, so we must use **rom** instead. Here is an example:

```
enum MESSAGE_LIST {MSG_PUMP_IDLE,MSG_PUMP_RUNNING,
               MSG_PUMP_DOWN,MSG_PUMP_OFF};
rom char rom * MESSAGE[]= {"Pump is on standby",
                           "Pump is running",
                           "Error detected in pump",
                           "Pump switch is in the OFF position"};
```

Making this a two-dimensional array is how you would have multiple lists for multiple languages. The above examples store the strings in program memory. For a lot of strings this takes a lot of space. Sometimes you will want to save the strings in an external device. For example a serial EEPROM. The extra time to read the strings is usually not a problem since humanly read displays don't change too fast compared to the processor speed.

Summary

- C strings are arrays of characters with the last character being a 0 or \000.
- **strlen()** counts the number of characters in the string before the terminator.
- **strcpy()** is a popular function to copy a string to another string.
- **strstr()** is one way to find one string inside another string.
- **strcmp()** is one of the key functions to compare one string to another and can easily be used to alphabetize.
- **strcat()** is used to concatenate one string to the end of another one.
- **gets()** and **puts()** are the traditional string input and output functions.
- The stream operators >> and << can be used in compilers that support the newer method for string input and output.

- **`atoi()`** and **`sprintf()`** are the traditional ways to convert from a string to a number and a number to a string.
- The stream operators >> and << can be used in compilers that support the newer method for string to integer conversion in both directions.
- C has a number of functions to determine what kind of character a character is.
- C characters are always 1 byte; however, Unicode characters are multi-byte and when used will require conversion by the programmer.

Exercise 12-1

Objective: Gain a working understanding of the standard C string-handling functions.
Requires: E3 module, USB cable, PC.

Steps/Technical Procedure	Notes
1. Write a program that will enter a string from the user, then convert all "can't" instances to "cannot" and all "won't" instances to "will not", then output the new string.	
2. Change Exercise 10-1 to instead of "int32 id_number" use "char name[10]" and to enter first names instead. For the list keep the names in alphabetical order, ignoring the case.	
3. Write a program that accepts commands and then acts on those commands. The program should display nice errors for improperly formatted commands. The commands to accept are: LED RED ON LED RED OFF LED GREEN ON LED GREEN OFF LED YELLOW ON LED YELLOW OFF READ location (location is a hex number, respond with data at that location in RAM) WAIT PRESS (wait for button press) DELAY n	

Quiz

(1) Which character cannot appear inside a C string?
  (a) A backslash \
  (b) A double quote "
  (c) An ASCII zero \000

(d) All of the above cannot appear inside a C string
(e) All of the above can appear inside a C string

(2) For the following code, what is the value of x?

```
char  s1[20] = "Hello";
char  s2[15];
int8  x;
strcpy(s2, s1);
x = strlen(s2);
```

(a) 20
(b) 15
(c) 5
(d) 6
(e) 0

(3) For the following code, what string is output?

```
char s1[20] = "This is a test";
char s2[] = "at";
strcpy(strstr(s1,"is"),s2);
cout << s1;
```

(a) This is a test
(b) That is a test
(c) This at a test
(d) That
(e) No way is this legal C

(4) For the following code, what string is output?

```
char s1[20] = "This is a test";
char * p;
p = strstr(s1,"t");
while(p!=0) {
    *p='s';
    p=strstr(s1,"t");
}
cout << s1;
```

(a) This is a test
(b) This is a sess
(c) This is a sess
(d) This is a s
(e) This is not valid C syntax

(5) For the following code, what string is output?

```
char s1[20] = "This is a test";
int8 x,i;
x = strstr(s1,"d")-strstr(s1," ");
for(i=0; i<x; i++)
  strstr(s1,"is")[-1]='?';
cout << s1;
```

(a) This is a test
(b) This ????? test
(c) Th????s a test
(d) T????is a test
(e) T??s??s a test

(6) For the following code, what string is output?

```
char s1[20] = "This is a test";
for(i=0; i<strlen(s1); i++)
   if( s1[i]<'n' )
      toupper(s[i]);
cout << s1;
```

(a) This is a test
(b) THIs Is A tEst
(c) THIS IS A TEST
(d) tHIs Is A tEst
(e) THis is a test

(7) What is wrong with the following code?

```
char s1[20] = "This is a test";
char s2[20] = "string";
strcat(s1,s2);
```

(a) The s2 declaration is too big
(b) The s1 declaration is not big enough
(c) There is no space between the strings
(d) The result of strcat() is not assigned to anything
(e) There is nothing wrong

(8) For the following code, what string is output?

```
char s1[20] = "This is a test";
char s2[20] = "string";
char s3[50];
strcpy(s3,"Why is ");
strcat(s3,s1);
strcat(s3,s2);
strcat(s3,"?");
cout << s3;
```

(a) Why is This a test string?
(b) Why is This a test string?
(c) Why is this a test string?
(d) Why is This a test string?
(e) Why is

(9) For the following code, what string is output?

```
int32  x = 123456789;
char s[20],s1[20],s2[20];
s << x;
for(signed int i=strlen(s)-3; i>0; i-=3) {
    strncpy(s1,s,i);
    s1[i]=0;
    strcpy(s2,&s[i]);
    strcpy(s,s1);
    strcat(s,",");
    strcat(s,s2);
}
cout << s;
```

(a) 123456789
(b) 1234,567,89
(c) 123456789,
(d) 123,456,789
(e) Nothing (the null string)

(10) For the following code, what is x?

```
int32  x;
char s[20]="000000";
for(int i=0; i<=4; i++) {
   s[i]='0'+i+1;
}
s >> x;
```

(a) 12345
(b) 0
(c) 000000
(d) 123450
(e) 12340

CHAPTER 13

Function-Like Macros

Macros are implemented in C with the *#define* preprocessor directive. The **#define**s used thus far are called *object-like macros*. *Function-like macros* are a variation of the standard **#define**. All the macro logic is done during the preprocessor stage and consists of text replacements. Unlike an object-like **#define**, the function-like replacement is done based on arguments supplied where the macro is used. Here is an example macro:

```
#define    INRANGE(var,lo,hi)     ((var>=lo) && (var<=hi))
```

It might be used like this:

```
if(  INRANGE( x, 10, 20 ) )
```

And this will be processed as:

```
if( (x>=10) && (x<=20) )
```

The three parameters (**var**, **lo**, **hi**) are identifiers whose scope is only on the **#define** line and the identifiers are replaced in the **#define** target text before the text replaces the macro invocation.

Macros may reference other **#define** identifiers and other macros. For example:

```
#define   HIGH_SUB(x)   (sizeof(x)/sizeof(x[0])-1)
#define   LAST(x)   x[HIGH_SUB(x)]
int16  table[20];
```

The following:

```
LAST(table) = 5;
```

is the same as:

```
table[19] = 5;
```

Notice the identifier x is used in both defines and these are different identifiers.

Embedded C Programming. http://dx.doi.org/10.1016/B978-0-12-801314-4.00013-2

Arguments

The arguments are always considered as text and each argument is separated from the next by a comma. The one special rule is if a comma is inside inner parens or a quoted string, then it is not an argument separator.

Here is an example that shows the power of the text argument:

```
#define  IF_RUNNING(stmt)  if( state==S_RUN && !fault ) stmt;
```

Used like this:

```
IF_RUNNING( output_high(PIN_B0) )
```

does this:

```
if( state==S_RUN && !fault ) output_high(PIN_B0) ;
```

Here is an example that shows how the comma rule is used:

```
   #define   MIN(x,y)   (x<y?x:y)
   #define   MAX(x,y)   (x>y?x:y)
   #define   INRANGE(var,lo,hi)     ((var>=lo) && (var<=hi))
   if( INRANGE( value, MIN( lowest, 5 ), MAX( highest, 150) ))
```

For INRANGE var is value

```
   lo is MIN( lowest, 5 )
   hi is MAX( highest, 150)
```

For MIN

```
   x is lowest
   y is 5
```

For MAX

```
   x is highest
   y is 150
```

The macro invocation may have fewer arguments than the macro definition. The missing arguments are treated as an empty string. The macro invocation may not have more arguments than the macro has parameters. Some compilers do issue a warning if there are too few arguments.

Macro Names

The macro identifiers are the same as any other identifier; however, there is one twist. The macro names with a (after them are treated separately from those without. That means you can have two macros with the same name, one with a (after and one without, as shown in this example:

```
#define MAX 31
#define MAX(x,y)   (x>y?x:y)
...
if( MAX( curent_data, highest_data) < MAX )
```

However, like many things in C, just because you can do it does not mean it is a good idea.

Concatenation Operator

For only macros, C has a concatenation operator, the **##**, that may be used to combine an argument value with an identifier or another argument value in the macro. Here is an example to show how it could be used:

```
char  input_buffer[100];
char  output_buffer[100];
#define CLEAR(name)  for(i=0;i<sizeof(name##_buffer);i++) \
                                    name##_buffer[i]=0;
...
CLEAR(input);
```

In the macro both **name##_buffer** become **input_buffer**. Note that you could not have used **name_buffer** because in this case the name would not be replaced with input because only full identifiers are matched and replaced.

We also show the \ preprocessor line continuation symbol to make a multi-line macro. Macros can get to be quite long so you will see this a lot with macros.

Stringize Operator

For only macros, C has a stringize operator; the **#** may be used to turn a text argument into a doubly quoted string. Here is a simple example:

```
#define  PIC(chip)  #include  #chip ".h"
...
PIC(16F887)
```

This unwraps to:

```
#include "16F887.h"
```

Notice the two strings "**16F887**" and "**.h**" are combined automatically by the preprocessor. The first **#** does not stringize because a parameter name does not follow it.

Here is a more complicated example combining both **#** and **##** from the C standard:

```
#define   str(s)   # s
#define   xstr(s)  str(s)
#define   INCFILE(n)  vers ## n
...
#include  xstr( INCFILE(2).h )
```

This results in:

```
#include  "vers2.h"
```

The spaces around the **#** and **##** are not included in the final string. At first glance it seems like the **xstr** macro is not needed. If we called **str** instead of **xstr** then the result would be:

```
#include  "INCFILE(2).h"
```

This is because the preprocessor first replaces the arguments and then re-processes the result to perform any further macro replacements. In the case where the **#** operator is involved, there is no macro replacement of the identifiers inside the double quotes. To do the replacements first and add the quotes afterward, we need two levels of macros as shown.

Here is a more practical example that shows how the same argument can be used inside and outside a string:

```
#define SHOW(x)   cout << #x " = " << x << endl;
...
SHOW(count);
```

will translate to:

```
cout << "count = " << count << endl;;
```

The image from a compilation that is loaded into program memory is a file with a **.hex** extension. The CCS C compiler allows the programmer to change the name of that file using the **#export** directive. The following is an example of a way to get the program version into the actual file name:

```
#define   MAJOR_VERS   3
#define   MINOR_VERS   1

#define STRZ(x)     #x
#define TOSTR(x)    STRZ(x)

#export(file="PUMPMASTER_" TOSTR(MAJOR_VERS) "_" \
        TOSTR (MINOR_VERS) ".hex", hex)
```

After a compile the final hex file will have the name:

PUMPMASTER_3_1.hex

Variadic Macro Syntax

Like for functions, newer C compilers permit the ellipsis syntax in a define macro. This is called a variadic and must always appear at the end of the parameter list. Because macros are evaluated at compile time, the functionality is somewhat limited for macros. The ... allows any number of arguments to be passed to the macro. All the arguments including the comma separators are then used as the replacement text for the special identifier **__VA_ARGS__** in the macro text. This syntax is frequently used with **printf()** because the number of arguments in **printf()** is variable. For example:

```
#define   PRINTIF( cond, ... )    if(cond) printf(__VA_ARGS__) \
          PRINTIF( x>250, "%lu is too many items", x );
```

Here is an example where function calls are easily redirected and the number of parameters varies:

```
#define CALL( function, ... )    DualUnit##function(...) \
        CALL( ClearQueue, 0,200,FALSE );
```

is the same as:

```
DualUnitClearQueue( 0,200,FALSE );
```

Here is another **printf()** example that directs the same output to two streams:

```
#define SEND(...) {fprintf(DGB,__VA_ARGS__);  \
                   fprintf(GPS,__VA_ARGS__);}

SEND("N%05lu,%05lu", x, y);
```

Function-Like Macros vs. Inline Functions

Macros and inline functions work very similarly. In both cases new code is generated at the point of invocation. The macros have an advantage of allowing for a greater variation in the generated code. For example, here is a macro to wait for a specified number of seconds but with an early exit condition specified by the caller. In this case we want to exit if the B0 pin goes low:

```
#define DELAY_SECS( time, abort ) for(int i=0; i<(time*10) && !(abort); i++) \
                                  delay_ms(100);
...
DELAY_SECS( 10, !input(PIN_B0) );
```

To do the same thing with an inline function you would first need to create a function to return the desired abort condition. You would then pass a pointer to that function to the inline **delay_secs()**.

Readability

Macros almost always make the code easier to read and review. Picking a good name for the macro will help a lot. Macros are commonly used for doing numeric conversions from one unit to another or to perform a common math computation. Consider as an example a program that internally deals with distance in inches; however, for setting certain limits it is easier to read and review as feet and inches.

```
#define FEET_INCHES(feet, inches)    ((feet)*12+inches)
#define UPPER_LIMIT                  FEET_INCHES( 15, 6 )
#define LOWER_LIMIT                  FEET_INCHES( 2, 3 )
#define MAX_MOVEMENT                 FEET_INCHES( 10, 0 )
#define MIN_MOVEMENT                 FEET_INCHES( 0, 2 )
```

The person who reviews this code may be well aware the upper limit is 15' 6" but would not as quickly recognize 186 as being the right number. It is very common that the units used in code do not correlate to the units used by humans. Macros can make the code more readable and easier to change.

Macros can be used to remove a lot of syntax in repetitive or frequently used code segments. Assume you have a structure like this:

```
struct {
   int16 id_number;
   int32 address;
   int8   type;
   int1   active;
}  device_list[20];
```

To assign values to one entry of this array of structures, a macro can be defined like this:

```
#define  SET_DEVICE( slot, id, addr, kind, act)  \
    {device_list[slot].id_number=id;  \
    device_list[slot].address=addr;  \
    device_list[slot].type=kind;  \
    device_list[slot].active=act;}
...
SET_DEVICE( 0, 0x23, 0x654332, 3, FALSE );
SET_DEVICE( 1, 0x19, 0x654112, 4, TRUE );
```

In this chapter all macros were shown in uppercase. Although not required, it helps to differentiate a macro call from a function call. Specific coding standards may specify the naming convention for a macro.

Advanced Example

Some embedded programs control products that have a critical security or safety component. Certifying organizations will have standards with coding requirements for these types of programs. One common requirement is that any critical variables be kept in two locations. Every time the variable is used both copies are checked to make sure they match. All writes are always to both copies. This is all done to make sure a hardware flaw does not corrupt a variable. This is done when the variable contents are very important.

An easy way to implement this requirement is to use a structure to keep both copies, like this:

```
struct {
   int32 encrypt_key;
   int32 local_id;
   char  security_level;
   int8  threat_count;
} secure_data_one, secure_data_two;
```

Then macros can be defined for write and read like this:

```
#define SECURE_READ(field)   ((secure_data_one.field==secure_data_two.field)? \
                               secure_data_one.field:flag_error(ERR_BAD_RAM))
#define SECURE_WRITE(field,data) {secure_data_one.field=data; \
        secure_data_two.field=data;}
```

The usage throughout the program would look like this:

```
SECURE_WRITE(security_level,'S');
if(SECURE_READ(threat_count)>2)
  shutdown();
```

Study the example so you fully understand how the macros work. Consider what the problems would be if this were attempted with functions instead of macros. Notice how easy it would be to change the program if the security requirements were to change. For example, three copies with majority vote.

Debugging Macro Problems

The *#warning* (or **#error**) directive may be used to find out how the compiler views a macro in some, but not all, compilers. For example:

```
#define  HIGH_SUB(x)   (sizeof(x)/sizeof(x[0])-1)
#define  LAST(x)   x[HIGH_SUB(x)]
#warning LAST(table)
```

will cause the following to appear in the error file:

```
>>> Warning 224 "main.C" Line 59 (9,21) :
    #warning table[  (sizeof (table) / sizeof (table [0])-1)]
```

Note the **sizeof()** expressions are not evaluated in the warning because those are not known to the preprocessor.

Summary

- Function-like macros have parameters and arguments like regular functions, except that the arguments are treated as text replacements in the macro.
- After the text replacement, the result is scanned again for further replacements until no more can be done.

- Commas not enclosed in inner parens separate the arguments in the macro invocation.
- Function-like macros and object-like macros may have the same name.
- The **##** concatenation operator can combine an argument with something else to make a new identifier.
- The **#** stringize operator will convert a text argument into a double-quoted string.

Exercise 13-1

Objective: Gain an understanding of how to use C macros.
Requires: E3 module, USB cable, PC.

Steps/Technical Procedure	Notes
1. Write a program that asks for a temperature in Fahrenheit and outputs the value in Celsius. To perform the conversion write a macro called Fahrenheit2Celsius.	
2. A program is being designed to control a plotter that has a 15-inch movement in the X and Y directions. A 16-bit integer is used to position the plotter such that 0 is at one end and 65535 is at the 15" mark. Write a program to draw a 5-inch square on the center of the plotter. This is done with two functions, moveto(x,y) and drawto(x,y). Each function takes an int16 for x and y. To test your code simply output to the screen the function name and the x and y values inside those functions. The main program should make the calls to draw the square using a macro that allows the positions in the code to appear in inches.	

3. Rewrite the following macro to no use macros but instead make a delay_secs() function. The second parameter will need to be a function pointer to a function that checks the desired abort condition (in this case B0 low).
`#define DELAY_SECS(time, abort)`
`for(int i=0; i<(time*10)\`
`&& !(abort); i++) \delay_ms(100);`
`...`
`DELAY_SECS(10, !input(PIN_B0));`	

Quiz

(1) The following macro shown earlier in this chapter will not work as intended under what circumstance(s)?

```
#define    MIN(x,y)   (x<y?x:y)
```

(a) If called with floating-point arguments
(b) If called with one floating-point and one integer argument
(c) If either argument is an expression
(d) If either argument has an expression with an assignment operator
(e) There are no problems with this macro

(2) What will the following warning show?

```
#define z              5
#define C              4
#define B(x,y)         x+C-y*z
#define A(x,y,z0+)     B(x,y)/z
#warning               A( B(1,2),3,0 )
```

(a) c+1+4−2*5
(b) 0+1+4−2*5+4−3*5
(c) 0+B(1,2)+4−3*4
(d) 0–16
(e) −16

(3) What will the following warning show?

```
#define A(a,b,c,d,e)   a|b|c|d|e
#warning    A( "ABC,DEF", (12,34), 9876)
```

(a) "ABC|DEF"|(12|34)|9876
(b) "ABCDEF"|(12|34)|9876|
(c) ABCDEF|(12|34)|9876|
(d) "ABC|DEF"|12,34|9876|
(e) "ABCDEF"|(12,34)|9876||

(4) What will the following warning show?

```
#define A(x)   (x)
#define B(x)   [x]
#define C(x)   {x}
#warning       C( B(1)+A(2) )
```

(a) C(B(1)+A(2))
(b) B(1)+A(2)
(c) {B(1)+A(2)}
(d) {3}
(e) {[1]+(2)}

(5) What will the following warning show?

```
#define A(x,y,z) =x=y=z=
#warning         A( 1,2 )
```

(a) 1,2
(b) =1=2=
(c) =1,2=
(d) =1=2==
(e) This will generate an error

(6) What will the following warning show?

```
#define          ONE   1
#define          TWO   2
#define          ONETWO 3
#define          THREE(a,b)   a ## b
#warning         THREE(ONE,TWO)
```

(a) ONE TWO
(b) 1 2
(c) 3
(d) ONETWO
(e) 12

(7) What will the following warning show?

```
#define   ONE   1
#define   TWO   2
#define ONETWO 3
#define FOUR(TWO)   TWO
#define THREE(a,b) FOUR(a) a ## TWO b
#warning  THREE(ONE,TWO)
```

(a) TWO 3 TWO
(b) TWO one 2 2

(c) 1 2 3
(d) 1 3 2
(e) This is not legal C

(8) If one wants to use a label inside a macro, how can the **##** be used to properly construct the label?
   (a) Use: **x##__line__**: where x changes for each label in the macro. The **__line__** is replaced with the line number, making a unique label
   (b) Use: **x##y**: where x changes with each label and y is a unique argument passed to the macro
   (c) Use: **x##rand()**: where x is unique for each label and rand is a function that returns a random integer number
   (d) Any of the above will work
   (e) None of the above will work

(9) What will the following code do?

```
#define X(s)   cout << #s << endl; s
...
X(table[12].first+=12);
```

   (a) Adds 12 to the member first in the 12th entry of table
   (b) Adds 24 to the member first in the 12th entry of table
   (c) Outputs to cout the member first in the 12th entry of table plus 12
   (d) Outputs to cout: table[12].first+=12
   (e) Outputs to cout: table[12].first+=12 and adds 12 to the member first in the 12th entry of table

(10) What will the following warning show?

```
#define     STR(s) #s
#define     ONE 1
#warning    STR(ONE)
```

   (a) ONE
   (b) 1
   (c) "ONE"
   (d) "1"
   (e) STR(ONE)

CHAPTER 14

Conditional Compilation

Conditional compilation was introduced briefly in Chapter 3. The directives associated with conditional compilation are preprocessor directives evaluated at compile time. These directives determine what code actually ends up getting compiled.

Consider an appliance that has three different models. One has a single LED, the second has five LEDs, and the third has an LCD screen. The core functionality of each is very similar, so you would like to maintain one code base that could be used to compile code for each model.

At the top of your project header file that is included by all the other files you might have a line like this:

```
#define MODEL  1000   // Change to 1000 or 5000 or 9000
```

And a sample function in the code might look like this:

```
void Show_Ready(void) {
   #if  MODEL==1000
      output_high( READY_LED );
   #elif  MODEL == 5000
      output_high( READY_LED );
      output_low( STARTING_LED );
      output_low( RUNNING_LED );
      output_low( WAITING_LED );
      output_low( ERROR_LED );
   #elif  MODEL == 9000
      lcd_putc("Ready");
   #else
      #error Invalid Model in Show_Ready()
   #endif
}
```

Certainly the above code could be written using standard **if** statements; however, in that case code would be generated for all three cases and the decision would be made at run time as to

Embedded C Programming. http://dx.doi.org/10.1016/B978-0-12-801314-4.00014-4

which path to take. Using the above code, ROM is used to save only the one case that is needed for the indicated model at compile time.

The model 1000 probably will fit into a much smaller PIC® chip, so you can also have something like this:

```
#if   MODEL==1000
   #include <16F683.h>
#elif   MODEL == 5000
   #include <16F876.h>
#elif   MODEL == 9000
   #include <16F1938.h>
#else
   #error Invalid Model in main.h file
#endif
```

After some time the hardware may change, and using conditional compilation to deal with that works out nicely. For example, say the 5000 model now has a rev 2:

```
#if   MODEL==1000
   #include <16F683.h>
#elif   MODEL == 5000
   #if REVISION==1
      #include <16F876.h>
   #else
      #include <16F886.h>
   #endif
#elif   MODEL == 9000
   #include <16F1938.h>
#else
   #error Invalid Model in main.h file
#endif
```

I/O pins may also be defined with conditional compilation, like this:

```
#if   MODEL==1000
   #define READY_LED   PIN_B0
#elif   MODEL == 5000
   #if REVISION==1
      #define READY_LED        PIN_B0
      #define STARTING_LED     PIN_B1
      #define RUNNING_LED      PIN_B2
      #define WAITING_LED      PIN_B3
      #define ERROR_LED        PIN_B4
```

```
   #else
      #define READY_LED          PIN_D0
      #define STARTING_LED       PIN_D2
      #define RUNNING_LED        PIN_D3
      #define WAITING_LED        PIN_D4
      #define ERROR_LED          PIN_D5
   #endif
#elif  MODEL == 9000
   #define LCD_PORT  "PORTD"
#else
   #error Invalid Model in main.h file
#endif
```

The above indentation helps to read the code; however, some programmers will not indent preprocessor conditionals. This is because some C compilers require the **#** to be the first character on the line.

Basic Directives

There are three flavors of the if directive:

#if pp-expression	TRUE if pp-expression is nonzero
#ifdef pp-symbol	TRUE if pp-symbol has been **#define**d
#ifndef pp-symbol	TRUE if pp-symbol has not been **#define**d.

The **#else**, **#elif**, and **#endif** apply to all of the if directives. The **#elif** is a combination of **#else** and **#if**; however, it does not need an extra **#endif** as would be needed if you used the two directives.

The conditional compilation directives affect other preprocessor directives as well as regular C code. Notice the **#error** in the above example. That line does not result in an error message unless the conditional compilation directive is satisfied. Code in the conditional compilation that is not in the TRUE area is treated just like comments. That means there could be C syntax errors and there would be no error message.

The **#ifdef** and **#ifndef** simply check to see if an identifier was **#define**d. It does not need to have a value assigned to be found, for example:

```
#define  DEBUG
#ifdef  DEBUG
    cout << "Value=" << value << endl;
#endif
```

This will output the diagnostic data.

Relational Expressions

The standard C relational operators as well as the parentheses are all valid in the expressions. The result is always evaluated as TRUE and FALSE. You can use numeric constants as well as char. String constants are not standard; however, some compilers allow them. The only variable names that can be used are names that were created with **#define**.

Special Macros

There are some special preprocessor macros that can be used in **#if**. The standard macro available is **defined()** and is used to avoid using **#ifdef** like this:

```
#if  defined(DEBUG)
```

This syntax can be useful because now other relational operators can be used. For example:

```
#if  defined(DEBUG) || defined(INTERNAL_BUILD)
```

This will be true if either of the identifiers is defined.

The CCS C compiler implements the **getenv()** as a preprocessor macro, so code like the following is allowed:

```
#if  getenv("PROGRAM_MEMORY")<32768
   #error This program needs a chip with at least 32K
#endif
#if  getenv("VERSION")!=5.123
   #warning This program was only tested with compiler version 5.123
#endif
```

Note that in the manual **getenv()** is in the built-in function section because that is where standard C defines it.

The CCS C compiler also has a preprocessor macro to help figure out how an identifier has been defined in C. The **definedinc()** macro returns a number to indicate if it is a local or global variable, a typedef, a structure, or anything else. 0 is returned if the id is not known to C. For example:

```
#if  definedinc(clear_string)==0
    void clear_string( char * string ) {
         *string=0;
    }
#endif
```

Special Defines

C has a few predefined defines that can be used as follows:

`__LINE__`	line number in the source code file
`__DATE__`	compile date (implementation-dependent format)
`__TIME__`	compile time (implementation-dependent format)
`__FILE__`	full source code file path and name.

The CCS C Compiler also has a few more:

`__DEVICE__`	full device name, like PIC16F887
`__FILENAME__`	just the filename, no path
`__PCx__`	defined for the current compiler (like **`__PCM__`**). The CCS compilers are named for each opcode family. For example, the high-end family (PIC18 parts) is PCH.

Global Defines

The technique described above where a common include file has the defines set up for the configuration is a popular method of dealing with multiple configurations. It does, however, require editing that file for each build. There is another method that most compilers support, although there is no standard method, called global defines. Many product builds are done in batch files executed on a PC. A typical command line looks like this:

```
ccsc  +FM  myproduct.c
```

This would compile the file and the resulting hex file would be made. The batch file might then go on to copy the hex file to a global location and maybe generate a log entry.

The following command line could be used to force some global defines throughout the whole project:

```
ccsc  +FM  #MODEL="5000"  #REVISION="2"  myproduct.c
```

Global defines can also be set in most IDEs. For example, in the CCS IDE use OPTIONS > PROJECT OPTIONS > GLOBAL DEFINES.

Strange Errors

Be aware that according to the rules of C if an identifier in an **`#if`** is not defined you do not get an error. Some compilers will give a warning. This can be frustrating if there is a spelling error in the identifier.

A missing **`#endif`** can cause the most bizarre errors. If missing in an include file, the error may not be in the include file itself. It just ignores the rest of that file and then starts ignoring code in the main file until it finds any **#endif**. From there, the errors now start in what seemed to be harmless code.

Be aware the **#if #endif** block can span across files.

Examples of Conditional Compilation

```
#define MAX_ID  500
#if  MAX_ID<256
   typedef  int8  ID_TYPE;
#elif  MAX_ID<65536
   typedef  int16  ID_TYPE;
#else
   typedef  int32  ID_TYPE;
#endif

#define SPEED_CONTROL_PIN PIN_C2

#ifdef  SPEED_CONTROL_PIN==PIN_C1
  setup_ccp2(CCP_PWM);
#elif  SPEED_CONTROL_PIN==PIN_C2
  setup_ccp1(CCP_PWM);
#elif  SPEED_CONTROL_PIN==PIN_B5
  setup_ccp3(CCP_PWM | CCP3_B5 );
#elif  SPEED_CONTROL_PIN==PIN_E0
  setup_ccp3(CCP_PWM | CCP3_E0 );
#else
  #error  Speed control pin is not a PWM pin
#endif

// Only even release versions contain diagnostics
#if ((VERSION%2) == 0)
   #define DEBUG
#endif

#if !defined(BYTE)
   #define  BYTE  int8
#endif

// This will take a while with a slow clock so restart the WDT if used
#if  getenv("CLOCK")<10000000  && getenv("FUSE_SET:WDT)
   restart_wdt();
#endif
```

Summary

- Conditional compilation is a powerful tool that can be used to have one set of source code files generate different configurations.
- Conditional compilation may be used to easily enable and hide diagnostic code in a program.
- **`#if`** can use only expressions that can be evaluated at preprocessor compile time.
- **`#ifdef`** and **`#ifndef`** simply check for the existence of a preprocessor identifier from a **`#define`**.
- The **`defined()`** and **`getenv()`** macros may be used to help make preprocessor decisions.
- Most compilers allow global defines to be set to establish a configuration just before compiling.

Exercise 14-1

Objective: Gain an understanding of conditional compilation.
Requires: E3 module, USB cable, PC.

Steps/Technical Procedure	Notes
1. Write a program that prompts the user for the width, length, and height of a room. Output the total volume of the room. The program should be designed to generate two versions, depending on the setting of a single define, to work in feet or meters.	

Quiz

(1) What does the following code display in the error file?

```
#define VERSION  3
#if VERSION>1
#if VERSION>2
   #error A
#else
   #error B
#endif
   #elif  VERSION==3
#error C
   #else
#error D
#endif
```

(a) A
(b) B
(c) C
(d) D
(e) This is not legal C

(2) What does the following code display in the error file?

```
#define NUMBER  1
#ifdef NUMBER
   #error Zero
#else
   #error One
#endif
```

(a) Zero
(b) One
(c) Zero and One
(d) Nothing
(e) This is not legal C

(3) Of the following, what is not a good use for conditional compilation?
(a) Maintaining multiple configurations of a project
(b) Using some of the same functions in many projects
(c) Enabling and disabling debug code
(d) Alerting the developer to illegal configurations
(e) These are all good uses

(4) What is output for the following code?

```
int8  XX = 1;
#define  XX 2
#if  XX==1
   cout << "ONE";
#elif XX==2
   cout << "TWO";
#else
   cout << "THREE";
#endif
```

(a) ONE
(b) TWO

(c) THREE
(d) Nothing
(e) There will be a syntax error

(5) What is output for the following code?

```
#define   XX 2
int8   XX = 1;
#if   XX==1
   cout << "ONE";
#elif XX==2
   cout << "TWO";
#else
   cout << "THREE";
#endif
```

(a) ONE
(b) TWO
(c) THREE
(d) Nothing
(e) There will be a syntax error

(6) Of the following applications, for which one is conditional compilation not going to solve the main objective?
(a) A printer with two ports, USB and Ethernet, where different code is used depending on which port is used
(b) A family of coffee makers where the code for each is similar but specific features vary
(c) An automobile computer where specific features change with each year's offering
(d) A phone system where certain customers get different features from everyone else
(e) All of these are good applications for conditional compilation

(7) Conditional compilation is great for dealing with different versions of hardware but why is it not good for maintaining different versions of software?
(a) Too much work to put in conditional directives for every change
(b) Rarely need to generate code for old versions
(c) Other tools like version control systems do a much better job of this
(d) The code would be too hard to read
(e) All of the above

(8) For the following code, what is output in the error file?

```
#define VERS "6.123"

#if strlen(VERS)!=5
   #error version length too long
#elif VERS[1]!='.'
   #error version format wrong
#elif VERS[0]>'q' ||VERS[0]<'0'
   #error major version must be 1-9
#endif
```

(a) Version length too long
(b) Version format wrong
(c) Major version must be 1–9
(d) No error message

(9) For the following code, what is output in the error file?

```
int8  junk = 10;
#if  junk<8
   #error junk too small
#elif  junk>10
   #error  junk is too big
#elif  junk==10
   #warning junk is just right
#else
   #error Something has gone wrong
#endif
```

(a) Junk too small
(b) Junk is too big
(c) Junk is just right
(d) Something has gone wrong
(e) This is not valid C

(10) What cannot be said about the following line:

```
cout << "Time: " << TIME << endl;
```

(a) It will display the current time to the user
(b) It will always generate different hex code
(c) It will show the time of the compile to the user
(d) There is no standard C way to format the time
(e) All of the above are valid statements

CHAPTER 15

PIC® Microcontroller

PIC® Architecture

In a simplistic way, Microchip's microcontroller line may be classified or categorized in the following groups:

- 12-bit opcode—baseline—10, 12, and 16 series part numbers
- 14-bit opcode—midrange—10, 12, 14, and 16 series part numbers
- 16-bit opcode—18 series part numbers
- 24-bit opcode—24, 30, and 33 series part numbers

Each group has its own specific capabilities. The number of bits in the instruction words corresponds to the opcode width (in bits).

The first three groups of microcontrollers above are classified as 8-bit devices because the data bus is 8 bits wide. This is the line to the right of the CPU in Figure 15.1. The left bus is 12, 14, 16, or 24 bits wide, depending on the opcode size. The last group above has a 16-bit-wide data bus. The PIC® uses what is called a Harvard architecture, where the program memory and data memory are on separate buses.

The PIC32 parts (32-bit opcode) use a very different architecture and are not covered in this book.

The processors come in packages as small as 6 pins or as large as 144 pins. For program memory the range is 512 instructions to 175,000 instructions. The RAM range is 16 bytes to 53,000 bytes. Some PIC® parts can operate as fast as 70 million instructions per second (70 MIPS). All these statistics are likely larger by the time you read this.

The PIC® processors are referred to as reduced instruction set computers (RISCs). The idea is there are only a small number of instructions, they perform simple operations, and most instructions are only one instruction word. The simplicity allows for fast execution and minimal logic (hardware gates) in the processor to operate.

The PIC16F887 part has 35 instructions and they are all single-word (14-bit) instructions. Most take a fixed four clock cycles to execute and branches take eight cycles.

Embedded C Programming. http://dx.doi.org/10.1016/B978-0-12-801314-4.00015-6

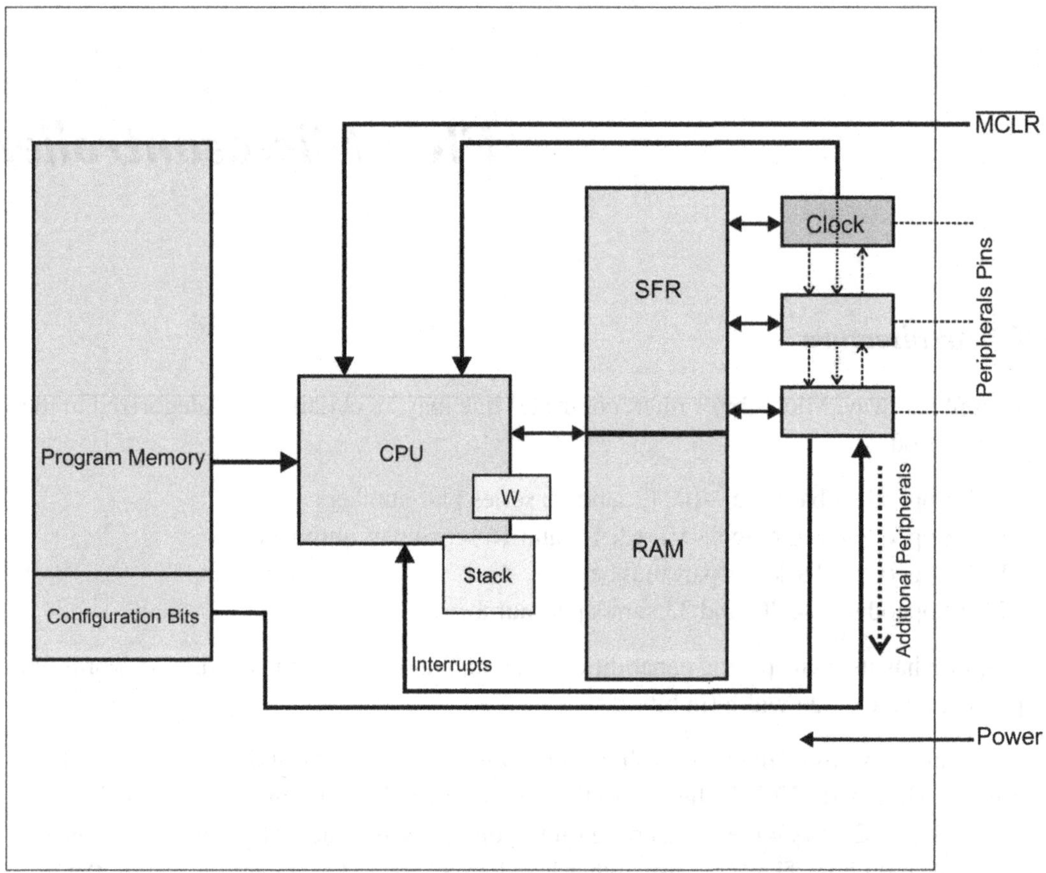

Figure 15.1: Typical PIC® MCU functional diagram.
W = working register.

In comparison, the processors in a desktop PC have instructions that can vary from 1 byte to 16 bytes. There are several hundred different instructions. Many can execute in a few clock cycles but some can take dozens of clocks. The PC has both program memory and RAM on the same bus; however, it does have a separate bus for what we call on the PIC® special function registers (SFRs).

CPU

Starting at the center of Figure 15.1 we find the central processing unit. The CPU will simply fetch an instruction from program memory (to the left) and execute that instruction. It then fetches another instruction, executes, and continues for as long as it has power and a clock, and the system is not in reset. On a PIC® the reset pin is called $\overline{\text{MCLR}}$ and it must be at a logic high level (like 5V) for the processor to run. The line over MCLR indicates for the action (master clear), the input must be low.

Stack

The CPU has access to a small area of memory called the stack. The stack was covered in Chapter 7 because on a PIC® the primary use is to implement function calls. When a CALL instruction is executed, the address to return to is pushed on to the stack. When that function issues another CALL instruction, another return address is pushed on the stack. When a RETURN instruction is executed, the top address on the stack is popped off and that tells the CPU where to return to.

Most C compilers also use the stack to save local variables. However, due to the small stack size on the PIC® and the lack of good instructions to access the stack, this is not practical. The stack is usually used only for CALL/RETURN on the PIC®.

PIC24 and up parts have the stack located in RAM as opposed to a hidden spot in the CPU. For these parts the stack may also be used to save temporary data and to do some parameter passing by the compiler.

For the other PIC® devices, the stack size varies from 2 to 32 locations.

Working Register

Figure 15.1 shows a single register called the *working register* (W). This is a single memory location that is special because most instructions in the processor have the ability to use the W register in the operations it performs. For example, there are instructions to add a number to the W register, to save the W register to a RAM location, or to load it from a RAM location.

On the PIC24 and up parts there are 15 W registers and they are located in the RAM. They are still special because the instructions can still operate with them by number. Even on those processors, the first W register `W0` is the most valuable because more instructions can use it than the others.

Special Function Registers

The RAM bus is split into two kinds of memory. One is general purpose memory such as what would be used for normal C variables. The other area has special memory locations that are used for special purposes called special function registers (SFRs). Many of the SFRs are connected directly to some peripheral device.

For example, consider a timer peripheral. The current value of the timer could be read by reading a specific SFR location. Another SFR location may be a control register for the timer where one bit of the register starts and stops the timer. The CPU (by way of the program in program memory) can read the timer value, change the timer value, or start and stop the timer by just reading and writing to the right memory locations.

The SFRs and RAM are on the same bus so they are equally easy to access from the processor and have different addresses.

Program Memory

Program memory is a non-volatile memory. All modern PIC® processors use a Flash memory technology that allows the program memory to be reprogrammed using a simple hardware interface. It is common to include some kind of programming connector on even a production product to allow for firmware updates if needed.

Some older PIC® parts that have a "C" after the first number are either one-time programmable or may have a window installed on the part to be UV-light erasable.

Normally a device programmer cannot only write to program memory but it can read the memory as well. All the PIC® chips have a configuration bit that can be used to read-protect the program memory so the device programmer can no longer read the memory. This may be done for production products so code cannot be stolen. Once the protect is enabled the only way a device programmer can access the chip is to erase the whole chip.

This is called code protection. Some device programmers also have an option to verify the program is secure. This does not prevent the chip from being erased and reprogrammed but it does prevent it from being read.

It should also be pointed out that for high volume projects, Microchip can preprogram the chips at the factory or you can have a parts distributer load the chips before delivery to you.

Instructions

An example of two instructions in a PIC16F887 part is shown in Figure 15.2.

Clock

The processor requires some kind of clock in order to run. The PIC® processors all have a wide variety of options for a clock source. A simple resistor and capacitor can be used if accuracy is not important. Commonly a crystal is used where the clock needs to be accurate. Many of the newer parts have internal oscillators that, although not as accurate as a crystal, are still very good. Several configurations are shown later in this chapter.

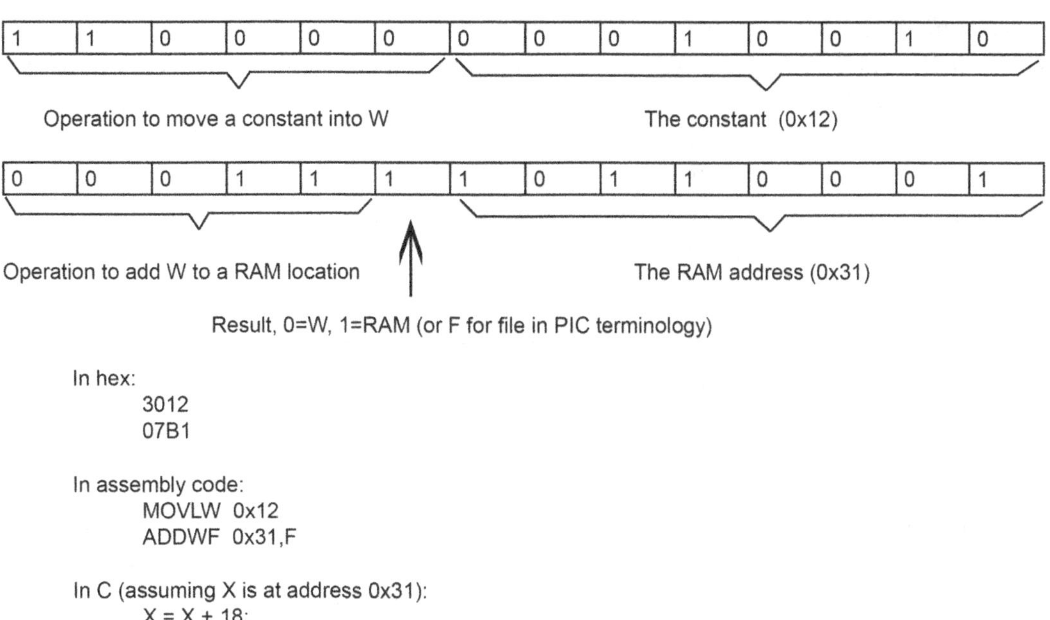
1 1 0 0 0 0 0 0 0 1 0 0 1 0
Operation to move a constant into W
The constant (0x12)
0 0 0 1 1 1 1 0 1 1 0 0 0 1
Operation to add W to a RAM location
The RAM address (0x31)
Result, 0=W, 1=RAM (or F for file in PIC terminology)
In hex:
3012
07B1
In assembly code:
MOVLW 0x12
ADDWF 0x31,F
In C (assuming X is at address 0x31):
X = X + 18;

Figure 15.2: Binary representation of two PIC® instructions.

The frequency of the clock controls the fetch/execute cycle of the CPU as well as providing a clock source to many of the peripheral modules. The PIC® CPU executes instructions in phases and most chips require four clock cycles to execute one instruction. For example, if you have a 20-MHz crystal this means the CPU executes 5 million instructions per second. The 5 MHz rate is referred to as the instruction clock. The 20 MHz is referred to in PIC® data sheets as *Fosc*. Some of the PIC24 parts work on a two-cycle clock for instruction execution.

Some PIC® devices have a built-in phase-locked loop (PLL) that can be programmed to multiply the external frequency. For example, you can use a 10-MHz crystal to generate a 40-MHz clock. In the CCS C compiler all the clock options can be set using the **#use delay** directive. The following is a directive for this example:

```
#use delay( crystal=10mhz, clock=40mhz )
```

The PIC® will pre-fetch the next instruction while it is still finishing execution of the current instruction. In the case that the current instruction requires a jump to another program memory location, then that pre-fetched instruction is discarded and a new fetch is needed. For this reason PIC® instructions that jump take twice as long (eight cycles) to execute that do all other instructions.

The internal clocks are calibrated at the factory. The way this is done varies depending on the chip. For some parts a calibration constant is saved at the end of program memory and the compiler must read that value and program it into an SFR to set the right clock rate. Most chips with an internal clock have some way for the running program to tweak the frequency by adjusting a value in an SFR. Over a normal operating temperature range, most data sheets indicate the accuracy of the internal clock is +/−2%. In reality the error might be half that. This is good enough for many applications. It is not good enough for a real-time clock, where a 2% error loses a half hour a day. RS-232 communication works with up to a 3% error so the internal clock works for that. USB needs a much more accurate clock with less than a 0.25% error. Usually this requires a crystal; however, some newer parts do have a high-accuracy internal clock.

Be aware most chips have a maximum frequency they can operate on dependent on the supply voltage to the chip. Higher voltages allow for higher frequencies.

Reset

When the processor starts running, most PIC® devices start at address 0 in program memory. This is called the reset vector. The C compiler will insert a jump here to the start of **`main()`**. The reason for a reset is kept in the SFRs and can be accessed in C using a call to **`restart_cause()`**.

Sleep

PIC® processors have a sleep feature that allows the chip to turn off the clock and stop executing instructions. This is usually done to save power when the chip has nothing to do. The sleep mode is activated by a special CPU instruction and from C a function call to **`sleep()`**. A wake-up from sleep can be done either by an interrupt such as a change on an I/O pin or by the watchdog timer going off and resetting the chip.

Interrupts

The peripherals have the capability to generate an interrupt to the CPU. An interrupt causes the processor to immediately call a function regardless of what it is doing. When that function finishes it returns to wherever it was when the interrupt happened. For example, an I/O pin can be configured to generate an interrupt when the pin changes from a low to high. If connected to a button then you could have a function called whenever the button is pressed. This all is covered in more detail in Chapter 17. For now, understand many of the peripherals have the capability to generate an interrupt to the CPU.

When the processor gets an interrupt trigger it uses an interrupt vector in program memory to determine where to go to. PIC16 and lower parts have a single vector for all interrupt types. The compiler inserts code to determine which interrupt happened and which function to call. PIC18 parts have two vectors (a high and a low priority) and PIC24 and up parts have one vector for every interrupt source.

Configuration Bits

Part of the non-volatile memory in the chip is used for what is referred to as configuration bits. This data is sometimes referred to as *configuration fuses*. These bits are used to control the peripherals; however, unlike the SFRs these bits are programmed with the program and stay through power cycles. As an example would be some bits that are set to determine what kind of clock oscillator is desired (crystal, RC, internal…). The clock peripheral then uses this to properly generate the clock on each power-up. In C these are set with the **#fuses** preprocessor directive. Many are automatically set by the compiler depending on what the program needs to do.

Peripherals

The boxes on the right-hand side of the functional diagram in Figure 15.1 represent the various peripheral modules in the processor. These vary a great deal between chips. Many will have physical connections to external pins on the chip. They also have access to the SFRs, clock, and configuration bits and can generate interrupts.

Table 15.1 lists all the peripherals as of this writing. The following chapters cover the most used peripherals in detail with the methods of use in C.

Minimal Hardware Connections

A neat characteristic of the PIC® microcontrollers is they require very few external components. Figure 15.3 shows how simple you can get by using a part with an internal oscillator. Some parts have multiple Vdd and/or Vss pins and it is good practice to put a bypass cap on each Vdd pin as close to the part as practical. Some parts also require another capacitor on a pin named Vcap. The required value will be in the data sheet. Some parts also have a configuration fuse that internally holds $\overline{\text{MCLR}}$ high. This saves a resistor for applications that do not require a reset pin. In the data sheet, pins are labeled with all possible uses. In addition to the port designation (port names are preceded by an "R" in the data sheet) will be the peripherals that can be connected to the pin. For example, a pin might be labeled like this: RB1/SDI/SDA. Some data sheets will instead list all the peripherals for each pin in a table, rather than showing them on the pinout diagram.

Device Programming

Chips may be programmed before being soldered to production boards or they may be programmed on the board. When doing the latter, some care must be used if the application uses the same pins as the programmer needs. The device programmer will use the $\overline{\text{MCLR}}$ pin (called Vpp for this application) to apply either a unique signal pattern or a special voltage level to put the chip into a special mode allowing the device programmer to read and write to

Table 15.1 PIC® peripheral list.

Ports	Most of the pins on the PIC® device are connected in groups of 8 (16 on the PIC24 and up) to what is referred to as a port. Ports have letters, and each pin a number starting at 0. Pin B4 is the 5th pin on port B. Programs can set these pins to a high or low state. The pins can also be used to input signals read by the program. Some pins can trigger an interrupt when they change state. This will all be covered in detail in Chapter 16. Many pins can be either routed to a port or to another peripheral depending on SFR or in some cases configuration bit settings
Timers	All PIC® devices have some number of units that can be used as a counter or timer. These will be covered in detail in Chapter 18
WDT	The watchdog timer unit was covered in Chapter 11. All PIC® microcontrollers have a watchdog timer; however the range of timeout times vary. Some chips have a maximum time of 2 s and others go up to several minutes
Data EE	Data EEPROM for user non-volatile data storage was covered in Chapter 11. Some PIC® devices have no data EEPROM and other have from 64 to 1024 bytes
ADC	The analog to digital converter allows the processor to sample a voltage at a pin of the PIC® and relate the voltage as a number to the program. Chapter 19 will go into more detail on the ADC
UART	The universal asynchronous receiver/transmitter unit allows for serial data transfer using a popular RS-232 protocol. This is underling hardware for **getc()** and **putc()** when connected to a PC serial port
SPI and I²C	Two- and three-wire serial buses are popular for communication between a processor and various hardware devices (for example a serial EEPROM). This is very short distance communication (usually on the same board). The PIC® has a module that will handle both the SPI and I²C standards. All this is covered in Chapter 20
PWM and CCP	Many PIC® devices have a module that is tied to the timers that can be used to generate pulses (pulse width modulator) or time external events (compare/capture). On many chips these are combined into a single unit and on others they are separate. In the PIC24 parts, they are called input compare and output compare. Chapter 21 will deal with these modules
Comparator	The comparator simply compares two analog voltages and generates an interrupt when one changes to be higher (or lower) than the other
Voltage detect	Some PIC® chips have the ability to perform some action when the voltage to the PIC® drops below a threshold or in some parts exceeds a threshold. Unlike the comparator, the voltage here is what is the chip is powered from. Some parts provide an interrupt and others only allow for a chip reset
Vref	Some PIC® devices have internal hardware that can generate a precision voltage that can be either output on a pin or routed to another internal peripheral. This can be used as a reference to properly scale or compare incoming analog voltages
Parallel port	Data can be transferred in parallel using 8 data pins and a strobe pin. This is the method used for older printers. This data transfer method can be used to transfer data very rapidly
RTC	A few PIC® chips have a built in real-time clock/calendar. The PIC® does require constant power and an external crystal for the clock. It can then keep time in hours, minutes, seconds...
USB	The USB bus is a serial protocol used to connect PCs to peripherals. Some PIC® microcontrollers have a module to handle this protocol. CCS has a separate development kit that includes a tutorial for this protocol

Table 15.1 Continued.

CANbus	The CAN bus is a serial multi-drop bus protocol that is popular in vehicles. Some PIC® devices have a module to send and receive data in this very specific format. CCS has a separate development kit that includes a tutorial for this protocol
Ethernet	Ethernet is used for local area network communication. Some PIC® chips have a hardware module to implement this popular protocol. CCS has a separate development kit that includes a tutorial for this protocol
Cap sense	A few PIC® microcontrollers have a module built in to sense capacitance changes on a pin. This is used for touch buttons—metal plates that can sense when a human finger touches them. CCS has a separate development kit that includes a tutorial for using this kind of human interface
QEI	It is common in motor control systems that there is feedback from an optical or magnetic encoder to indicate motor movement. The quadrature encoder interface module can be used to track these movements automatically so a position is tracked by the QEI module
DAC	The opposite of an ADC, the digital to analog converter allows the processor to output a specific analog voltage. Not many processors have this capability
Ext PM	A few PIC® devices allow the expansion of program memory to an external memory device. This requires a lot of pins for the hookup but will allow for programs that need a lot of memory. This is usually needed when there is a very large data table (like a dictionary), however can be used for very large programs as well
DSP	The digital signal processor is actually an extension to the CPU, not a true peripheral. It adds new instructions that specialize in complex math operations that are common for vector arithmetic. This allows for faster computing when doing signal processing such as digital filtering or voice analysis
DCI	The DCI module in some PIC® chips allows the sending and receiving of digitized audio using a popular standard format
DMA	Many of the PIC24 class parts have a direct memory access (DMA) module that acts like an interface between RAM and some other peripherals. For example, a DMA module can be programmed to automatically take all data that comes in from the serial UART and put it into a buffer in RAM with no effort from the program
CRC	A few PIC® devices have built-in hardware to perform the required math to calculate a CRC. This allows for much faster processing for applications that need to do a lot of CRC calculation
CWG	The complimentary waveform generator on a few processors produces two complementary PWM signals that can be used for motor control
NCO	A numerically controlled oscillator on a few processors is a fancy counter/timer module that increments by values other than the one a standard timer/counter uses
CLC	The configurable logic cell on a few processors runs a number of I/O pins into a programmable logic unit that can be set up by the program to implement certain logic in the hardware to allow decision-making at speeds much higher than the firmware can do
DSM	The digital signal modulator takes several PWM signals and produces a single signal out. It is like a carrier wave modulated by data. Only a few PIC24 class parts have this
Op amps	A few PIC® microcontrollers have built-in operational amplifiers

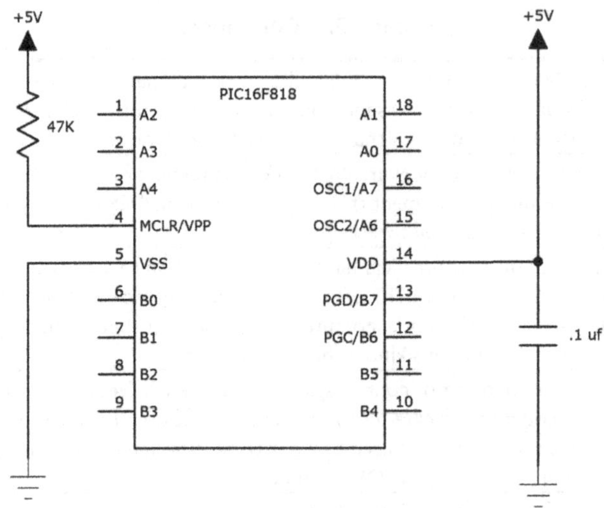

Figure 15.3: Example minimal PIC® MCU schematic.

the program memory (and configuration bits). Two I/O pins are used, called PGC and PGD. Figure 15.4 shows an updated schematic showing a typical programming connector.

The device programmer needs only a five-wire connection to the target board. Frequently a sixth wire is added to be used for debugging. Device programmers simply load a hex file produced from the compiler into the chip. The following is a list of the different kinds of device programmers:

- **Chip programmer:** Will have a zero-insertion-force (ZIF) socket to insert a chip to be programmed
- **Gang programmer:** Like the above but will have multiple sockets to program many chips at the same time
- **In-circuit programmer (ICSP):** Connects directly to a target board to program the chip on the board. Sometimes the board must be powered and sometimes the device programmer provides the power. A common PIC® tool for this is an ICD unit. An ICD unit can also be used for debugging
- **In-circuit gang programmer:** Like the above, but will program multiple target boards at the same time
- **Remote in-circuit programmer:** This is a hand-held unit that is first loaded with the hex file and then the battery-operated unit is used to program the target boards away from the PC.

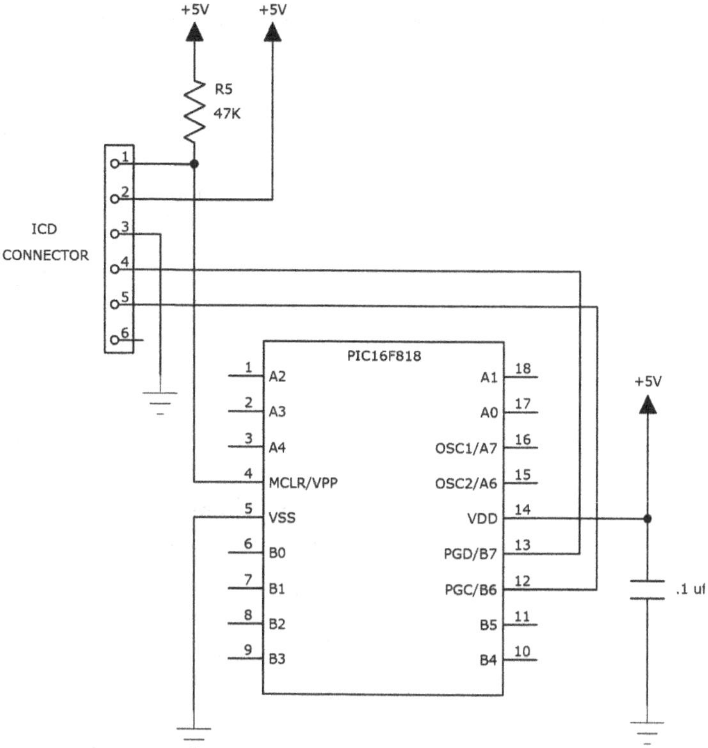

Figure 15.4: Example ICSP schematic.

In addition to the loss of two I/O pins, usually a programming/debugging port in the product is an extra expense because of the cost of a connector. There are ways around this however. Card edge connectors can usually be made with no additional expense. There is also a growing standard for no-connector connectors among PIC® developers, called Tag-Connect. The Tag-Connect is a pogo-stick contact device that presses down on to pads made on the PCB using a few guide holes. The PCB needs only some holes and pads to support the Tag-Connect. More information on this is on the CCS web site.

Hex Files

The standard hex file format used for PIC® devices uses an old Intel standard. These are referred to as Intel hex files. These are text files you can edit in Windows with Notepad. The only lines that are used in the hex file are lines that start with a : character. A **;** character is frequently used to indicate a comment line in the hex file. A typical line in a hex file is detailed in Table 15.2.

Table 15.2 Interpretation of Intel hex file format.

:1000400083050313A001A101E2010130E1000230A8	
:	Intel hex indicator
10	Number of bytes of data on this line
0040	Address of this line of data should be put into memory. This address is always a byte address, not the more frequently used instruction or word address
00	Line type: 00 is data, 01 means no more data in file, 04 is used to supply the upper 16 bits of the address for future lines
8305 ... 0230	Data
A8	Checksum for the line

Power-Up Considerations

Most PIC® microcontrollers have a power-up timer that will hold the device in reset for a period of time giving the power supply time to stabilize. Other processors use a capacitor on the reset pin to achieve a similar effect. The power-up timer can usually be enabled or disabled by a configuration bit. This is important because if the power is intermittent when it starts up, thousands of instructions may execute and then the chip goes down and restarts. This can cause an undesired effect.

Another problem is microprocessors can act a bit insane when operated below the minimum voltage. Another PIC® feature that can be used to combat this problem is the brownout detector. Again, with configuration bits a voltage level can be set, below which the processor is automatically held in reset.

Clock Configurations

The diagrams in Figure 15.5 show the various ways to clock a PIC® chip. Be aware all PIC® chips have a maximum frequency at which they can operate. Check the data sheet for your part to make sure you do not exceed the maximum clock. Some parts have a second slower limit for the speed of an external clock (like a crystal). Be aware the faster the clock, the more power the chip consumes. Sometimes a battery-operated unit will switch clocks at run time from a slow, low-power clock to a fast, accurate clock only when needed.

Debugging

Most of the newer PIC® microcontrollers have built-in debug capability. A debugger allows a program to be executed until a specific address is reached and then it will stop, allow the user to look at RAM, and execute one instruction at a time (single step) to go again until another address is reached. An in-circuit debugger (ICD) unit is required to do this kind of debugging.

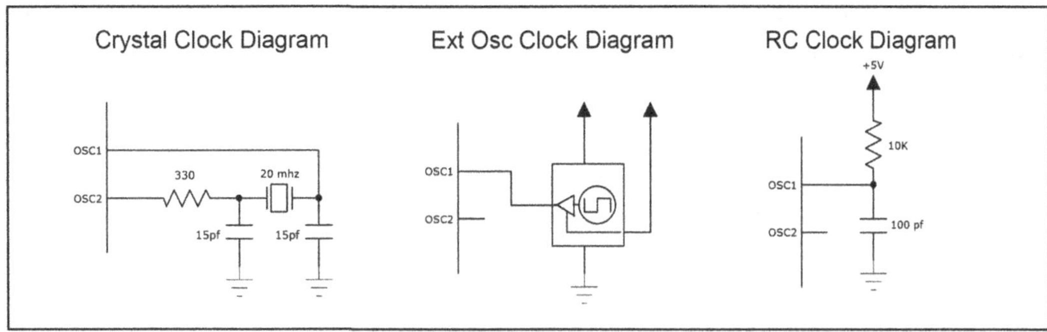

Figure 15.5: PIC® oscillator schematics.

It works with a configuration bit that indicates the chip should run in debug mode. When in debug mode, on reset the processor jumps to the end of memory, where a special debug program should have been loaded with the normal program. That debug program will communicate over the PGC and PGD pins to the ICD unit. When the ICD unit wants the user program to run, it will send a command to the debug program and it will jump to the user program.

A special register in the chip can be set to an address, and when the processor reaches that address it jumps back to the debugger program. Whenever the debugger program has control it can allow the ICD unit to read or write to RAM locations.

Some simple hardware additions to the processor along with a small debugger program, the ICD unit, and software at the PC provide some very powerful debugging capability.

The drawbacks for this kind of debugging are as follows:

- The PGC and PGD pins cannot be used by the application.
- Some program memory must be dedicated to the debugger (like 256 instructions).
- A few RAM locations must be dedicated to the debugger program.

A few PIC® devices with a low pin count do not use two of their valuable pins for debugging. Instead they have a special version of the chip with more pins (part numbers have an -ICD suffix) that can be used for debugging. These are usually put on a small board called an adapter, and that board connects to the ICD and to the socket on the target board where the processor usually resides. If you are using a low-pin-count chip you may want to make sure you have what you need for debugging. As an alternative, some will do most of the debugging with a larger chip and do a new layout for the production boards.

Bootloading

The primary method to get a program into program memory is to use a device programmer as described above. An alternate method to write a program to program memory is to use a bootloader. The key to a bootloader working is that most PIC® microcontrollers allow the program running in the chip to write to program memory. The following is a typical bootloader configuration:

The first 512 instructions in the chip are dedicated to a bootloader program. This program is loaded with a standard device programmer.

On reset, the bootloader program starts running and checks a "loaded" flag in non-volatile memory.

- If the flag is TRUE then the bootloader jumps to the application program starting at the 513th instruction.
- If the flag is FALSE the bootloader program waits for data over some communications interface. For example, RS-232, USB, or I²C. Once data is received it is written to program memory. After all data is written the loaded flag is set to TRUE and the chip is reset.

If an interrupt comes in, the bootloader may get it because it is in low memory, it will jump to the application interrupt handler so the application runs normally.

If the application program wants to update itself, it clears the loaded flag and forces a reset. This causes the bootloader to activate and load a new program.

Bootloaders are sometimes used in production products. For example, some TV sets allow you to plug in a USB Flash drive and the TV firmware, if it recognizes a file with the right name on the drive, will download it into program memory. You also will see service tools that connect up to industrial equipment having the ability to download new firmware through the RS-232 or USB port the service tool uses. Devices connected to the internet can be set up to bootload over the internet to get updates, much like many PC programs do. This is common on LAN routers.

Summary

- PIC®s are grouped according to the instruction opcode size: 12 bits to 32 bits.
- PIC®s vary based on program memory size, RAM size: and the peripherals offered in each chip.
- The peripherals may have external pins, access to specific SFRs, and use some configuration bits.
- All PIC®s have a watchdog timer and a clock.
- Internally, in addition to the various peripherals, the chips have program memory, RAM, special function registers, a stack, configuration bits, and one or more working registers.
- The PIC® has a Harvard architecture for the memory organization, separating the program memory and RAM on separate buses.
- The PIC® has a RISC instruction set that on some parts can execute 70 MIPS.
- The stack on most PIC®s is used for holding the return address for function calls.

- The clock can be configured for many different sources such as RC, crystal, or internal.
- In-circuit debugging is possible with only two I/O pins and a small amount of program memory and RAM using an ICD unit.
- Bootloading is a method of loading a new program into the chip without a device programmer.

Exercise 15-1

Objective: Become familiar with all the various PIC® device hardware features.
Requires: E3 module, USB cable, PC.

Steps/Technical Procedure	Notes
1. Using the `getenv()` options, write a program that will fully describe the PIC® chip the program is compiled for. Information should include the number of program memory locations, RAM locations, and EEPROM locations, and what peripherals are present in the chip	

Quiz

(1) A PIC® rated at 10 MIPS will not actually execute 10 million instructions each second for what reason?
  (a) The processor only executes one instruction for every four clock cycles
  (b) Programs are not always doing stuff, sometimes the code is idle
  (c) The program speed depends on the voltage to the chip and the voltage is never an exact amount
  (d) Some instructions take twice as long to execute
  (e) None of the above; 10 MIPS means 10 MIPS

(2) If instead of master clear ($\overline{\textbf{MCLR}}$) the chip designers wanted to call the pin RUN then the nomenclature for the pin would be what?
  (a) $\overline{\textbf{RUN}}$
  (b) RUN
  (c) !RUN
  (d) $\overline{\text{MRUN}}$
  (e) $\overline{\text{MCLR}}$

(3) Say a typical part has 8K of program memory and 1K of RAM. The opcode size is 14 bits and the RAM width is 8 bits; then what is the minimum bit width of each entry in the stack?
  (a) 8 bits
  (b) 10 bits
  (c) 13 bits
  (d) 14 bits
  (e) 32 bits

(4) If a chip designer decided to use eight working registers instead of one, then this would have an impact on what major part of the design?
   (a) He will need eight times the RAM
   (b) He will need more peripheral modules
   (c) The chip will require a higher voltage
   (d) The clock will need to run faster
   (e) There will need to be more bits in each instruction

(5) Of the following things that can be selected, what is not a good use of the configuration bits?
   (a) Selecting what pin a peripheral uses
   (b) Selecting the kind of clock the PIC® should use
   (c) Enabling or disabling the $\overline{\text{MCLR}}$ pin
   (d) Protecting program memory from being read by a device programmer
   (e) Setting an I/O pin to be high or low

(6) Which of the following applications are not good applications for an internal clock?
   (a) Set-back thermostat, that allows the user to program the temperature set-points for day and night
   (b) Sump pump controller that enables the motor for 30 s when a sensor so indicates, but there must be a 60-s cool-down period between activations
   (c) Blender controller that controls motor speed based on five user settings
   (d) Toy that blinks LEDs and makes noises when a button is pressed
   (e) These are all good internal clock applications

(7) Assume a project that uses a crystal clock, and has an LCD unit that needs seven I/O pins, three push buttons, and four LEDs. What is the minimum number of pins you will need on a PIC® for the project?
   (a) 7
   (b) 9
   (c) 14
   (d) 16
   (e) 18

(8) The data EEPROM is not shown on the processor block diagram on either memory bus. What is the most likely reason for this?
   (a) The memory is a part of the special function registers, just not separately identified
   (b) The memory is in one of the peripheral blocks, accessed by SFR
   (c) The memory is buried inside the CPU, accessible only to CPU instructions
   (d) The graphic artist missed the block
   (e) The data EEPROM is not a part of the processor chip

(9) In designing a project, one of the design goals is to have special code running in the processor used only during production testing. After that different code is run exclusively to implement the end-user functionality. Of the following ideas for implementing this goal, which idea will not work?

(a) Put a programming jack on the board and, using a device programmer, program in the production test program, after testing load in the user program

(b) Using a chip large enough for both test and user function, preprogram the chips with one program with both functions. After the production test passes, a flag is set in EEPROM, preventing the production test function from ever being called

(c) Preprogram the chips with a bootloader and the production test program. Use some port on the board, like RS-232 or USB, to program the user program via the bootloader after production testing

(d) Preprogram the chips with the production test program in program memory and load the user program into RAM. After the production test is complete, the production program copies the RAM program to program memory

(e) All of the above methods will work

(10) For question 9, which method is the most expensive per unit?

(a) All of the methods will have a similar cost

CHAPTER 16

Discrete Input and Output

We have already been using some basic discrete input and output using the basic built-in functions *input()*, *output_low()*, and *output_high()*. These functions operate on a single pin. Additional pin-related functions are as follows:

output_bit(pin, value);	sets a pin to low or high based on value (0 or 1)
output_float(pin);	sets a pin to high impedance (input, no drive)
output_drive(pin);	sets a pin to output (high or low dependent on the last state set)
output_toggle(pin);	switches the pin state from high to low or low to high
value = input_state(pin);	if the port is set to input the return is the same as input(pin). If output the value is the last state output.

Pins are grouped into ports and we have the following functions to operate on an entire port. In each case value is a bitmap where the least significant bit of value goes to pin x0, the next bit to pin x1, and so on.

output_x(value); x is the port letter like A or B

value = input_x();

Input Voltages

When we talk about I/O in terms of low and high or 0 and 1, that is a digital view of the world. From an electrical view these pins have a specific voltage. The PIC® pins have two types of digital inputs: TTL and Schmitt trigger. The data sheet will identify which pins are of which type. For a TTL pin on a part with 5 V, like the Vdd, the typical rules are as follows:

0–0.8 V reads as low or 0
2–5 V reads as high or 1.

There is no consistency for what the reading will be between 0.8 and 2 V. For digital input pins the PIC®s very much want a voltage near Vss (0 V) or Vdd (5 V in this example). Supplying a voltage midrange can cause the chip to draw a higher current.

Embedded C Programming. http://dx.doi.org/10.1016/B978-0-12-801314-4.00016-8

The trigger points vary depending on the Vdd. The data sheet will have details for each specific chip. For example, when Vdd is 3.3 V the following may be the rules:

0–0.5 V reads as low or 0
1.6–3.3 V reads as high or 1.

Some input pins on the PIC® may be designated as Schmitt trigger inputs. These inputs never change digital state in the no-man's-land between the thresholds. For example, once the voltage is low enough for a digital 0 it will not change to a digital 1 until the upper threshold is reached. This builds in some hysteresis, preventing noise on a signal from causing the digital values to rapidly change. The Schmitt trigger thresholds are a little higher than TTL. For example, see Figure 16.1.

Digital circuitry has traditionally been 5 V. There is a slow trend toward using 3.3 V instead. Many modern designs have a mix of 3.3-V and 5-V parts. From the above data you can see a part that outputs 0 V and 3.3 V can be read by a 5-V part yielding a correct 0 and 1. The other direction is more of a problem since normally 5 V on an input can damage a 3.3-V part. Some PIC® devices have some pins designated as 5 V safe when running at 3.3 V. Make sure to check the data sheet for your device to find out if that feature is available and on which pins.

Drive Current

On a 5-V Vdd part the **`output_low()`** and **`output_high()`** will have ideal output voltages of 0 V and 5 V. On output what is more of a factor is the current the pin is able to supply. Many PIC® parts can supply up to 20 mA (0.02 amps) for both 0 V and 5 V.

When it supplies 5 V it is referred to as sourcing current, and for 0 V it is called sinking current. A 20 mA source and sink current is considered very good. Some pins on some PIC® chips may be designated as having a lower source and/or sink capability. Even though each pin may be able to supply 20 mA, there is still a limit to what all pins together can source from a single chip. This means that at any given time only a few pins can drive a full 20 mA.

To better understand the implications of drive current we will use the following equations from Ohm's law:

Current = voltage/resistance
Resistance = voltage/current
Voltage = resistance * current

5 V	
0 to 1 V	reads low
2.05 to 5 V	reads high

3.3 V	
0 to 0.66 V	reads low
2.6 to 3.3 V	reads high

Figure 16.1: Schmitt trigger thresholds.

Consider an imaginary light bulb with an internal resistance of 500 ohms connected to a pin of a 5-V part (see Figure 16.2). With an **output_high()** call the pin becomes 5 V and with the 500-ohm bulb there is a 10-mA flow of current. With an **output_low()** the pin is 0 V and the other side of the bulb is 0 V so there is no current flow.

In reality even a small incandescent bulb is less than 80 ohms and will draw too much current for a PIC® device. LEDs are a favorite light for PIC® chips. An LED is a form of semiconductor diode and requires the current to be limited. An LED has a fixed forward voltage drop (Vf) and maximum forward current (If). Find the smallest resistor to use with the formula (Vdd – Vf)/If. This will provide maximum brightness without burning out the LED. Larger resistors will dim the LED. For example, an LED with a Vf = 1.7 V and If = 10 mA will use a minimum resistor of (5 – 1.7)/0.010 = 330 ohms in a 5-V system. Figure 16.3 is an example to drive an 18-mA-maximum LED with 15 mA (Vdd = 5V).

LEDs have polarity. The line on the LED is the cathode and needs to be connected to ground while the other pin (the anode) is connected to a positive voltage to light. The example in Figure 16.3, is is configured so an **output_high()** lights the LED. To light the LED with an **output_low()** use as shown in Figure 16.4.

Some LEDs have a built in resistor; however, that makes the LED specific to a certain voltage.

Driving More Current

Sometimes you do need more than the 20 mA the PIC® can drive. Relays, for example, almost always need more current than a PIC® can provide. What is generally done is a transistor is used to drive the real load. The PIC® drives the transistor at a much lower current. Figure 16.5 shows a typical configuration for an 80-ohm light bulb. The transistor sinks 63 mA to the bulb and draws only 2 mA from the PIC®.

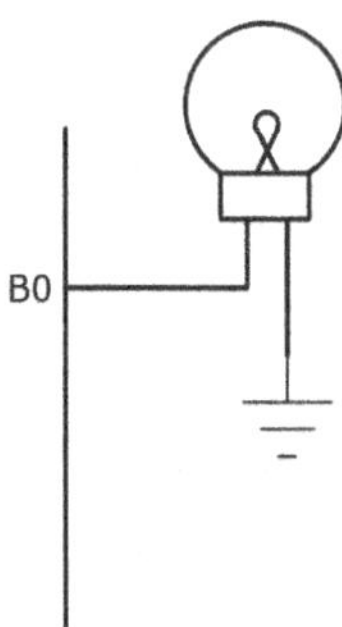

Figure 16.2: Light bulb connected to a PIC® MCU.

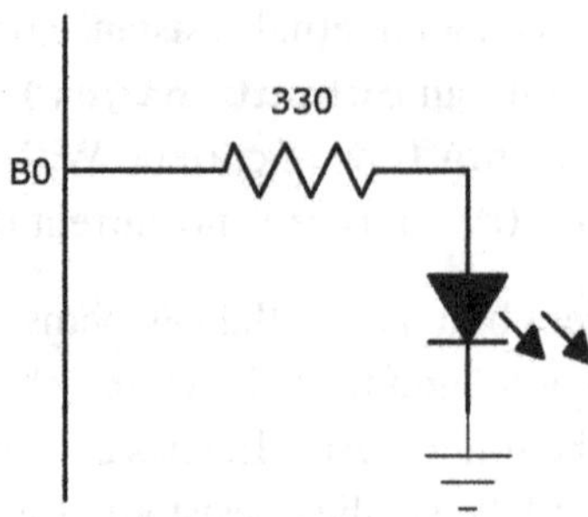

Figure 16.3: LED connected to a PIC® MCU.

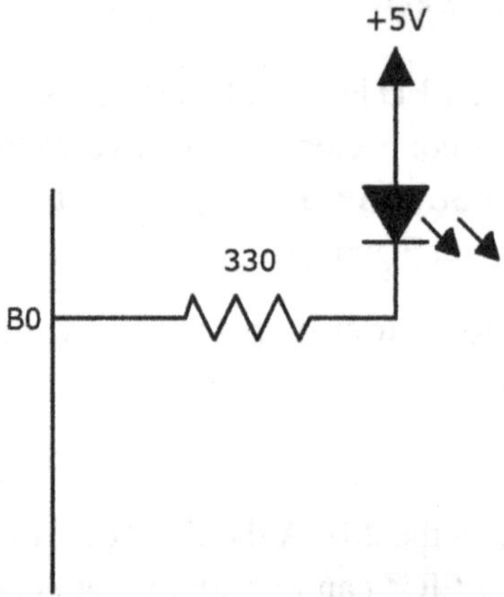

Figure 16.4: Negative logic LED connected to a PIC® MCU.

Open Collector Outputs

Some pins on the PIC® may be designated as open collector outputs. This is sometimes called open drain. On most PIC®s the A4 pin is open collector. What this means is the pin can sink current but not source current. The effect is an **`output_low()`** operates normally and an **`output_high()`** does the same thing as making the pin an input. One use of this is when multiple devices are connected to the same wire with a pull-up resistor. Any device can do an **`output_high()`** or **`output_low()`** without fear of one device sourcing when another device sinks causing a large current flow in both parts. The signal will be low if any device tries to pull it low. The signal will be high if all devices try to set it high (due to the pull-up resistor).

This configuration is a kind of hardware **`or`** function. One example of use would be as an error flag. If any processor in a multi-processor system has a critical error, it pulls the pin low. Each processor knows that one or more of the other processors is in error if the pin is low.

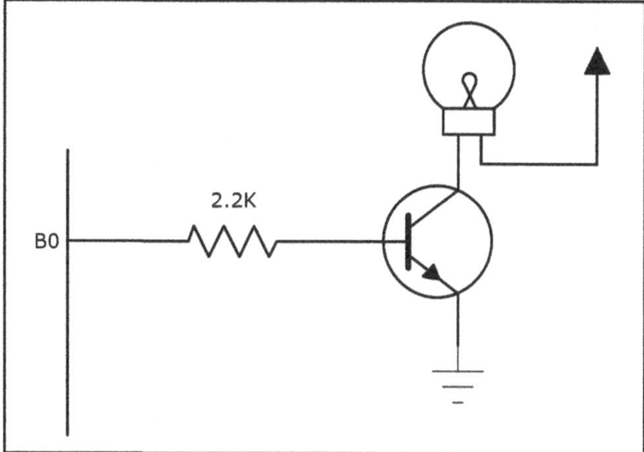

Figure 16.5: NPN transistor driver schematic.

There is no harm if multiple processors signal an error or if some processors do not.

This is not commonly used, but it is very common for someone to report pin A4 is not working correctly on their chip because they cannot drive it high.

Direction

Each digital pin can be configured as an input or output. Internally this is done with a special function register (SFR) called the *tri-state (TRIS) register*. By default the compiler sets the direction to output when an output function is used (like **`output_high()`**) and sets it to input when an input function is used (like **`input()`**). This is referred to as the *STANDARD_IO mode*.

The drawback to this automatic setting of the direction is that extra code (and time) is taken on each I/O operation. There is an advantage in that the programmer does not need to keep track of the direction and some chips that are susceptible to ESD may actually change direction in response to overvoltage pulses. By frequently refreshing the TRIS register, the correct direction is maintained. There can also be an advantage to slowing down the outputs a bit. This allows the pin to reach the desired voltage before the next operation.

For those that want to directly control the direction (TRIS register) the *FAST_IO* mode can be used. The entire port is put into the **`fast_io`** mode like this:

```
#use fast_io(B)
```

That puts port B into the **`fast_io`** mode from this point in the file until the mode is changed. Now the programmer must manually set the direction like this:

```
set_port_b_tris( 0x0F );
```

This sets pins B0 to B3 as input (the 1 bits) and pins B4 to B7 as outputs (0 bits). An **output_high(PIN_B0)** will not do anything; the B0 pin remains an input. In **fast_io** mode most output operations happen in one instruction. This is very fast.

When using **fast_io** mode, the operators are sometimes too fast. Consider the following code:

```
output_high(PIN_B0);
output_low(PIN_B1);
```

If B0 does not reach the TTL high voltage before B1 is set low, then it will be set low with B1 on some PIC® devices. Some PIC® chips read the port, modify the correct bit, and write it back. This read–modify–write can cause other pins in the port to change if they are not read correctly.

There is a third I/O mode for those who want to fix the direction of each pin to a certain direction but also want to refresh the TRIS on each I/O operation. This is the *FIXED_IO mode.* When setting the **fixed_io** mode you must specify each output pin, and the remaining pins on the port are set to inputs. For example:

```
#use fixed_io(b_outputs=PIN_B4,PIN_B5,PIN_B6,PIN_B7)
```

Button Input

A very typical input is from a push button. The most common type of button has only a single make/break (single throw) so you cannot use it to directly set a pin to either 0V or 5V. To solve the problem, a pull-up resistor is generally used to get the 5V and the switch is used to get the 0V. Figure 16.6 shows a typical connection.

Pull-Ups

The diagram in Figure 16.6 shows a weak pull-up resistor being used to apply 5V at a low current to the PIC® pin. Since the PIC® is looking at voltage, the low current doesn't matter. Pull-up resistors are commonly used on input pins. Many PIC® parts have the capability to enable internal pull-up resistors on selected pins. This eliminates the need for the extra part(s). There are several different ways this feature is implemented, depending on the part. Some chips only offer this feature on one port (usually B) and only allow you to turn all the pull-ups on or all off. The call looks like this:

```
port_b_pullups(TRUE);
```

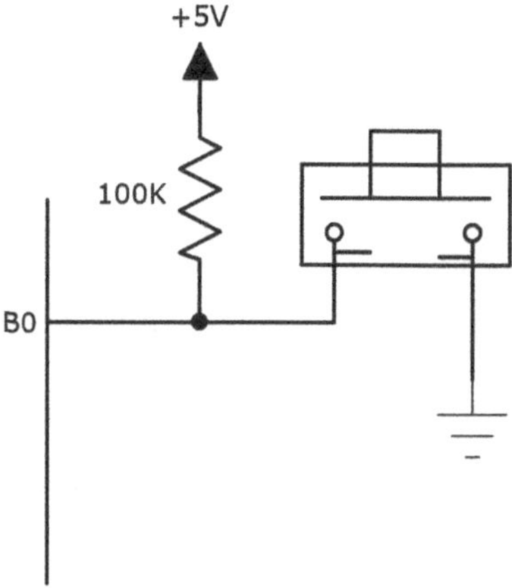

Figure 16.6: Push button connected to a PIC® MCU.

Other parts allow you to specify a bitmap like the **set_tris_b()** function. For example:

```
port_b_pullups( 0x03 );
```

turns pull-ups on for pins B0 and B1 only. For the PIC24 parts the pins are treated separately so the function call looks like this:

```
set_pullup(TRUE, PIN_B1);
```

A small number of parts also have pull-down resistors that can be enabled. For these we use the same function and pass two masks, one for pull-ups and the second for pull-downs:

```
port_b_pullups(0x03, 0x04);
```

This does pull-ups on B0 and B1 and a pull-down on B2.

Debounce

A switch and many other signal sources do not always make a clean make/break. Figure 16.7 gives a scope trace showing a button press that could easily be interpreted by the super-fast firmware on a PIC® as multiple presses.

For example, consider the following code to detect a button press and release:

```
while(input(PIN_B0)) ;    // Wait until a press
while(!input(PIN_B0)) ;   // Wait until a release
```

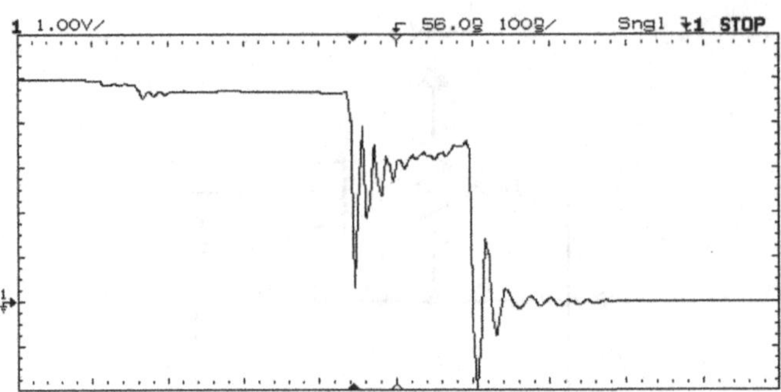

Figure 16.7: Scope trace of a button press.

This code will assume there is a release due to the noisy press.

The fix is to debounce the input. For example, the following will ignore any noise for the first 50 ms after a press:

```
while(input(PIN_B0)) ;        // Wait until a press
delay_ms(50);
while(!input(PIN_B0)) ;       // Wait until a release
```

There are many methods of doing the debounce and it will depend on the overall logic of the program to implement the best method.

Filtering

The above example shows how to filter out noise at a specific spot where noise is expected. Some systems have more consistent noise that needs to be ignored by the firmware. It is very important to understand the noise you have in order to properly filter it out. A scope will be a huge help in this effort. Consider a signal where you are expecting to do something in response to a 5-ms pulse. Figure 16.8 gives a scope trace that shows the noise that must be filtered out.

We see the good pulse; however, you can also see lots of problem pulses. One characteristic of the problem pulses is they are always less than 10 us wide. Here is a filter to ignore pulses less than 20 us:

```
do {
   while(!input(PIN_B0)) ;
   delay_us(20);
} while(!input(PIN_B0));
```

This takes two samples 20 us apart and they must both be high to accept the pulse. If the noise happens to be 20 us apart this wouldn't work. If the noise is very irregular then you may not

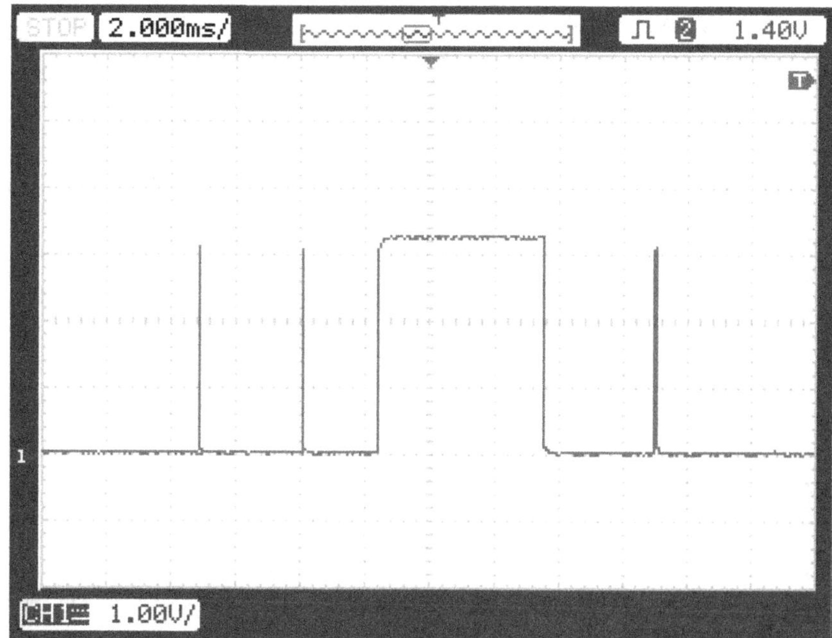

Figure 16.8: Scope trace of a good pulse with noise.

want to rely on this method. In that case you may want to take a number of samples and make sure they are all high. The scope trace is critical in designing a good filter.

Although the above method works well for occasional fast pulses, it would not work well for noise like that shown in Figure 16.9.

This data is from a radio receiver. The desired pulse is 600 us. There are lots of other pulses sent by the radio. Say we use a method like the above where instead of 20 us we use 500 us to ignore pulses less than 500 us. The problem is the second pulse and third pulse are 500 us apart and might be seen as a 500-us-long pulse. For this kind of noise a timer should be used and the check should be that the signal remains high for the full time. Timers will be covered in Chapter 18.

Memory-Mapping Ports

Sometimes a programmer will want to directly access the port without using the built-in functions. The following is a simple example:

```
#bit  LED1 = getenv("SFR:PORTB").2
...
LED1 = 1;
```

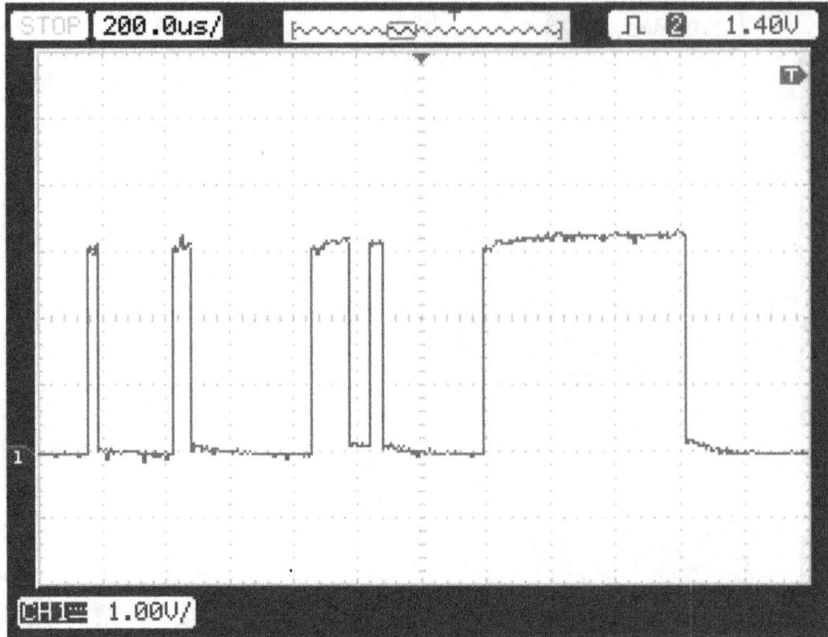

Figure 16.9: Scope trace of a good pulse over noisy RF link.

This sets B2 to a 1. Note that nothing is done with the direction (TRIS) register in this case. In the following example an entire structure is placed on a port:

```
struct {
   int unused : 1;
   int enable : 1;
   int read_write : 1;
   int data_control : 1;
   int data : 4;
} lcd_unit;
#locate  lcd_unit = getenv("SFR:PORTB")
...
lcd_unit.data = 5;
```

This sets B4 and B6 to a 1. B5 and B7 are set to 0. All other pins on B are unchanged.

Another way to do the same thing is:

```
output_b( (input_b() & 0x0F) | (5 << 4) );
```

Summary

- For the input of digital signals the voltage levels of the source and trigger levels of the PIC® must be considered.
- PIC® input pins offer standard TTL levels and Schmitt trigger inputs.
- Sometimes digital inputs require some filtering or debouncing.
- Digital outputs must be checked for the load they are driving to make sure the PIC® can handle it.
- In addition to checking each output pin, consideration must be taken for all pins combined.
- Most output pins are standard sink/source but some may be designated as open collector outputs.
- To drive more current than a pin can handle a driver transistor may be used.
- The direction of each pin can be handled automatically by the compiler or manually by the programmer.
- Some PIC®s offer internal pull-up resistor options on some pins that may be enabled or disabled by the program.
- It is important to understand the signal quality and account for noise and bounce in the program's logic.

Exercise 16-1

Objective: Learn how to perform discrete input and output operating in C. Experience with both seven-segment LED and keypad.
Requires: E3 module, USB cable, PC, 3 x 4 keypad, seven-segment LED with resistors on each anode and wires to connect to the E3.

Steps/Technical Procedure	Notes
1. Write a program that will repeat the following actions: a. Light the green LED and wait for a button press b. After the button is released, light just the yellow LED for 1 second c. Then light just the red LED for 1 second d. Finally turn off all LEDs and sound the buzzer with a 1-khz tone for 3 seconds. The buzzer on this board is actually just a piezo speaker. To make a tone you must generate a 0 to 5V square wave to the device at the desired frequency. **HINTS:** • Consider what must be done regarding debounce of the switch. • Make sure the tone sounds right and is not half or double the desired frequency.	

(continued)

Steps/Technical Procedure	Notes
2. Assemble hardware to connect a seven-segment LED module to the PIC® and design a program to count from 0 to 9 and then repeat. Each digit should display for a second (see Figure 16.10). The following is how the seven-segment LED is configured. Segments are labeled a–g and the decimal point is DP. Each segment is a separate LED with the cathodes connected together. **HINTS:** • Connect all LED pins to a single port and use the output_x(value) function to set all eight pins to the desired pattern. This means you must find a port on your board that has all eight pins coming out to the terminal block. • Construct a constant lookup table with the correct bit patterns to form each digit. For example, LOOKUP[3] would have the right value to show a 3 on the LED.g debounce of the switch.	
3. Assemble hardware to connect a 3 × 4 keypad to the PIC® and write a program that will display at the PC each key that is pressed on the keypad. Figure 16.11 shows how the keypad is configured. When a key is pressed a connection is made from one of the row pins to one of the column pins. For example, when the 4 button is pressed pin 5 is shorted to pin 2. The PIC® used in the E3 board allows internal pull-ups to be enabled only on port B and each pin can be set on or off individually. **HINTS:** • The PIC® is much faster than the human hand so the technique to use is to scan the keypad by setting one of the column pins low, the other columns high, and then check each row pin for one that is low. If none are found switch to another column and repeat. • For this to work each row pin must have a pull-up. Make sure to connect the row pins to a port on your chip that can enable internal pull-up. • Once a low is found on a row pin then use that row number and the current column number to index a constant array that translates the row and column to a character (like "4").	

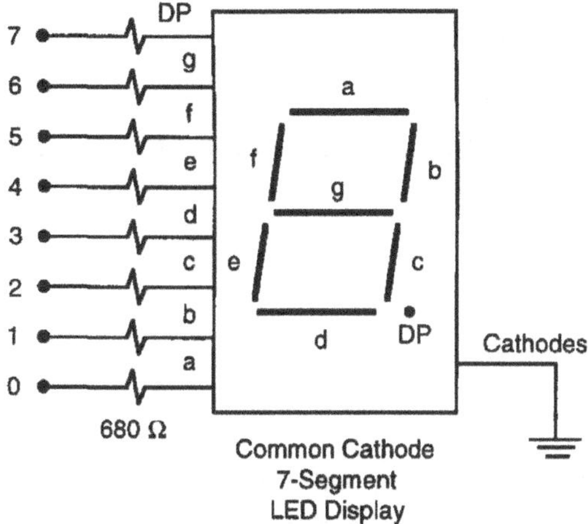

Figure 16.10: Seven-segment LED configuration.

	5	6	7
1 —	1	2	3
2 —	4	5	6
3 —	7	8	9
4 —	*	0	#

Figure 16.11: 3 x 4 keypad configuration.

Quiz

(1) If an open collector output pin is needed and there is none available on the PIC® then which two functions can be used to achieve the same effect?

(a) **`output_high()`** and **`output_low()`**
(b) **`output_float()`** and **`output_low()`**
(c) **`output_bit()`** and **`output_low()`**
(d) **`output_high()`** and **`output_float()`**
(e) **`input()`** and **`input_state()`**

(2) An LED rated at a maximum of 5 mA should use what value resistor in series to a 5-V output pin?

(a) 1 ohm
(b) 5 ohms

(c) 0.001 ohms
(d) 1000 ohms
(e) This LED won't work with a PIC®

(3) A relay with a coil rated at 100 mA can be connected to a 5-V PIC® using what extra component?

(a) 5-ohm resistor
(b) 50-ohm resistor
(c) 500-ohm resistor
(d) 5000-ohm resistor
(e) Transistor

(4) A seven-segment LED is connected directly to a PIC®. Each segment draws 20 mA. The PIC® can supply each with 20 mA with a maximum total drive current limit of 110 mA. The application does not need all the digits. It needs only count from 1 up for a short way. How high will it be able to count to?

(a) 3
(b) 5
(c) 6
(d) 7
(e) 9

(5) The following code outputs a pulse stream on pin B0. The while loop gets implemented as a single jump instruction at the bottom of the loop. Each output takes four instructions. For a PIC18 class chip with Fosc = 40 mHz, what will be the frequency of the square wave?

```
while(TRUE) {
    output_high(PIN_B0);
    output_low(PIN_B0);
}
```

(a) 1 mHz
(b) 1.111111 mHz
(c) 4 mHz
(d) 4.444444 mHz
(e) 40 mHz

(6) On a 5-V PIC® with a 1.5-V battery connected to one input pin, what is the digital value that will be read by **input()** on that pin?

(a) 0
(b) 1

(c) 1.5
(d) Depends on whether the input is TTL or Schmitt trigger
(e) There is no way to know

(7) Why doesn't the following code work as the programmer intended?

```
output_high(PIN_B0);
if(!input(PIN_B0))
   cout << "unable to drive pin high";
```

(a) There is not a delay between the output and input
(b) There should not be a **!** before the **input()**
(c) The **cout** lines needs to be inside **{** and **}**
(d) The I/O direction changes with the **input()** call
(e) The code will work as intended

(8) For a 3 × 4 keypad connected directly to a PIC® with no internal pull-ups, what is the lowest number of external resistors that will be needed?

(a) 1
(b) 3
(c) 4
(d) 7
(e) 8

(9) A program needs to detect a button press on a wire that picks up a lot of noise with 1-ms pulses. Of the following, which is the best method to ensure a real press?

(a) When a press is detected, wait 1 ms and read it again to make sure it is real
(b) Take four samples 500 us apart and require all four to be the same
(c) After the first detect wait 100 ms and make sure the button is still pressed
(d) Take four samples 25 ms apart and require all four to be the same
(e) Take four samples with no delay between and require all four to be the same

(10) The following code waits for an incoming pulse on pin B0. The while loop gets implemented as two instructions (a compare and a jump). The input takes two instructions. For a PIC18 class chip with Fosc = 20 mHz, what is the fastest pulse this code can be guaranteed to detect?

```
while(!input(PIN_B0)) ;
```

(a) 50 ns
(b) 200 ns
(c) 800 ns
(d) 1 us
(e) This code will detect any size pulse

CHAPTER 17

Interrupts

When an event occurs that demands the microcontroller's attention, an interrupt may be generated that will suspend what it is doing, take care of the task that needs to be performed, and go back to what it was doing.

As shown in Figure 17.1, when an interrupt occurs, the instruction currently being executed is completed. Then the PIC® jumps to the interrupt vector in program memory and executes the instruction stored there. Compiler-generated code may cause (as required) the microcontroller to first take notes on the status of the program it was executing when the interrupt occurred (context saving) so that it can find its place when it comes back. Then the interrupt service routine will handle the interrupt by doing whatever needs to be done. On completion, the routine will set everything back to the way it was and return to the main program where it left off. The code needed to both preserve and restore the status of the program that was being executed when the interrupt occurred is created automatically by the compiler and is called context saving.

Interrupts are caused by events that must be dealt with at the time they occur. Interrupts may come from several hardware sources.

Simple Interrupt Example

For all PIC®s that support interrupts (and most do) there is always one or more generic I/O pin designated as an interrupt pin. These pins simply will cause an interrupt when they change state as requested by the programmer. The hardware pins are identified as INTx where x is nothing or a number 1–4). In the compiler we call this INT_EXTx (EXT for external to the chip). There are three things that must be done in the program to handle an interrupt.

- Define a function to handle the interrupt when it happens using **`#int_`**... before the function.
- Enable the specific interrupt in the processor using the **`enable_interrupts()`** function.
- Turn on interrupt processing in the chip using **`enable_interrupts(GLOBAL)`**.

Embedded C Programming. http://dx.doi.org/10.1016/B978-0-12-801314-4.00017-X

Here is a full example program:

```
#include <E3.h>
int count;

#int_ext    // On the E3 the INT pin is B0
void button_isr(void) {
    count++;
}

void main(void) {
   count = 0;
   enable_interrupts(INT_EXT_H2L); // Enable EXT interrupt when B0 goes
                                   //  from 1 to 0
   enable_interrupts(GLOBAL);      // Start interrupt processing
   while(TRUE) {
      delay_ms(1000);
      count << "We got " << count << " interrupts." << endl;
   }
}
```

This program starts by showing a count of 0 every second. Once the push button is pressed it starts outputting 1, and when pressed again 2, and so on. We will use this simple program to explore some of the more complex interrupt techniques in the following sections.

Where Does the Time Go?

Assume we put a **delay_ms(1000)** inside the interrupt function (usually called an interrupt service routine or ISR for short). If the program runs for 10 seconds and the user presses the button three times, how many count lines are output? The answer is around seven. That is because for each button press the program was suspended for a second. The delay function in **main()** has no knowledge of the time lost in the ISR. From the user's perspective the data is output each second except when the button is pressed, and then there is a 2-second gap.

Consider this ISR:

```
#int_ext
void myisr(void) {
   while(!input(PIN_B0)) ;
}
```

Here we get the interrupt when the button is pressed and B0 goes low. We then stay in the ISR as long as the button is down. This has the effect of freezing the program as long as the user's finger is on the button. As you can imagine, except in very well planned out circumstances, delays and while loops of this nature are bad for use in an ISR.

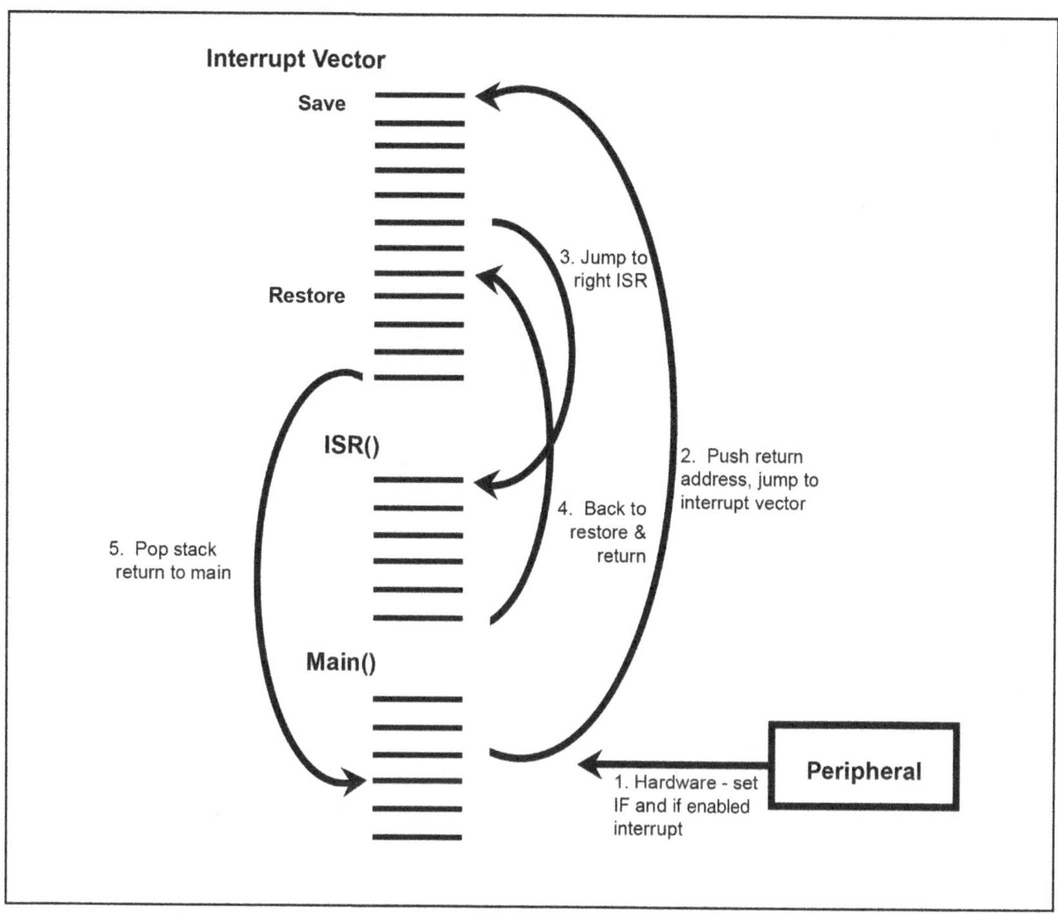

Figure 17.1: Interrupt sequencing diagram.
IF = interrupt flag.

Debounce Revisited

By now you realize a switch press can generate noise. When using interrupts each noise pulse can generate an interrupt. This can cause all sorts of trouble. An interrupt handler cannot interrupt itself so as long as you are in the interrupt handler another interrupt will not happen. Therefore the following code, similar to what we previously did, would work:

```
#int_ext
void button_isr(void) {
   count++;
   delay_ms(100):
}
```

However, it is generally a bad idea to delay inside an ISR, so we need another method. One way to solve the problem looks like this:

```
#include <E3.h>
int count;
int1 got_ext=FALSE;

#int_ext    // On the E3 the INT pin is B0
void button_isr(void) {
   got_ext=TRUE;
}
void main(void) {
   count = 0;
   enable_interrupts(INT_EXT_H2L);         // Enable EXT interrupt when B0
                                           // goes from 1 to 0
   enable_interrupts(GLOBAL);              // Start interrupt processing
   while(TRUE) {
      delay_ms(1000);
      if(got_ext) {
         count++;
         delay_ms(100);
         got_ext=FALSE;
      }
      count << "We got " << count << " interrupts." << endl;
   }
}
```

For this program all this work is a bit silly, but this technique can work well whenever you have too much (or too long) work to do in an ISR. Knowing the interrupt happened is often enough and then what needs to be done can be done when convenient. Notice that the way this code was written it will only count one press every second.

For simple debounce or filtering in an ISR it is more common to use a timer to figure out how much time passes between interrupts and to ignore interrupts that come in too close to one another. This will be covered in more detail in detail in Chapter 18.

It's Not Always a Good Time to Interrupt

The following program shows a very common problem programmers encounter with interrupts. The goal of the program is to capture the state of the B4 and C1 pins when B0 goes low and to make sure they are the same. The main program will output any errors.

```
#include <E3.h>

int1 B4_state, C1_state =FALSE;

#int_ext
void button_isr(void) {
   B4_state=input(PIN_B4);
   C1_state=input(PIN_C1);
}

void main(void) {
   count = 0;
   enable_interrupts(INT_EXT_H2L);    // Enable EXT interrupt when B0 goes
                                      // from 1 to 0
   enable_interrupts(GLOBAL);         // Start interrupt processing
   while(TRUE) {
      if(B4_state!=C1_state)
         count << "Error, When B0=0 then B4=" << B4_state << " and C1=" <<
            C1_state << endl;
   }
}
```

When dealing with interrupts you must always consider what would happen at any possible point the interrupt could happen. Here is the problem scenario for the above:

- Interrupt happens and saves 1 for both states
- Main program loads **B4_state** for the compare
- Before the main program loads the **C1_state** an interrupt happens
- The ISR saves 0 for both states this time
- After a return to the main the **C1_state** is loaded and the compare is done
- The compare shows **B4_state** as 1 (it was before the second ISR) and **C1_state** as 0 (it was after the second ISR)
- An error is displayed to the user when in fact B4 and C1 were never different.

One solution would be to do the comparison in the ISR and save an error flag and error data for the main program. This would work for this example, but there are many more complex situations that would be impractical to do in the ISR.

Another method is to stop interrupt processing while you deal with data that might change in an ISR. A simple modification looks like this:

```
disable_interrupts(GLOBAL);
error=B4_state!=C1_state;
if(error)
   count << "Error, When B0=0 then B4=" << B4_state << " and C1=" <<
      C1_state << endl;
enable interrupts(GLOBAL);
```

If an interrupt is generated while the global interrupts are disabled then that ISR is queued up to execute when interrupts are re-enabled. However, if two interrupts from the same source come in while interrupts are disabled then the ISR is invoked only once when interrupts are re-enabled. This is the most popular technique for solving this kind of problem; however, if this were more than an example we would want to code it such that the **cout** was done after interrupts were enabled. It is best to keep interrupts disabled for as short a time as possible.

The above example shows the problem with two variables. We would not have the problem with a single 1-byte variable. However, multi-byte variables can cause a problem. For example, in the ISR if you do this:

```
myint32 = 0x12345678;
```

and in the main program you do this:

```
if(myint32==0)
```

then we have the same problem. The first byte of **myint32** is loaded and checked for 0, then before the second byte is loaded and checked the ISR comes in and writes to all bytes, giving main an inconsistent view of **myint32**.

Why Do We Need Interrupts?

The alternative to interrupts is frequently checking for some condition throughout your program. For example:

```
distance = (sensor_value*23)/(cal_value+3);
if(!input(PIN_B0))
   output_high(GREEN_LED);
lcd_putc << "DISTANCE=" << distance;
if(!input(PIN_B0))
   output_high(GREEN_LED);
if(distance>99)
   change_range(HIGH_RANGE);
if(!input(PIN_B0))
   output_high(GREEN_LED);
```

You can imagine the trouble with this, especially if you have a number of different asynchronous events (events not directly related in time to the program execution sequence).

What is Really Happening

Interrupt Flag (IF)

It all starts with a bit called the interrupt flag (IF). There is one flag for every interrupt source. The flag is set to a 1 by hardware gates when the interrupt is detected. There is a huge advantage to the hardware detecting the interrupt because it can detect conditions too fast for firmware in the part to catch. For example, if you have a loop that is waiting for a pin to go high, while the jump at the bottom of the loop is executing you could have a single fast pulse and, by the time you do the input at the top of the loop, the pulse is gone.

The built-in function **interrupt_active(INT_EXT)** will return the state of the interrupt flag. The interrupt flag is always set when the peripheral detects an interrupt condition. Interrupts do not need to be enabled for the flag to be set. In fact some programs may not need an ISR, they just poll the IF to see if it was set. If another interrupt happens while the IF is set then nothing happens. If IF is 1 and it gets set to 1 again, nobody notices.

The interrupt flag is automatically cleared at the end of an ISR by the compiler. Sometimes the programmer will want to clear the flag manually. For example, the IF may have been set before a programmer enables interrupts and handling the pre-interrupt is not desirable. In that case the following is done:

```
clear_interrupt(INT_EXT);
enable_interrupts(INT_EXT_H2L);
```

Another case is when you are processing an interrupt and there might be another interrupt happening while you are in the ISR that you don't want to lose. In that case you do a **clear_interrupt(INT_EXT)** at the start of the ISR so the flag can be set again. In this case, however, you must tell the compiler not to clear the interrupt at the end. This is done like this:

```
#INT_EXT  NOCLEAR
```

Remember, the second interrupt will not be serviced until the first ISR completes.

The IF can be set as well to force a call to the ISR. In the compiler this is done by simply making a normal function call to the ISR function. The compiler will just generate code to set the IF for the function. When the IF flag is set, the interrupt function, of course, will not be called if the IE is not set. It will be called, however, as soon as the IE is set.

Interrupt Enable Flag (IE)

There is another flag for every interrupt to indicate if the interrupt is enabled. This is the flag affected by **enabled_interrupts()** and **disable_interrupts()**. A given ISR will not be called if the IE bit is clear.

Global Interrupt Enable Flag (GIE)

The global flag is used to indicate if the processor should handle any interrupts. This flag is set by **enable_interrupts(GLOBAL)**. No ISR will be called if the GIE bit is clear.

Interrupt Handling

If the GIE is set, a specific IE is set, and the corresponding IF is set, then the processor interrupts program execution. For 12–16-bit opcode parts this means the processor jumps to a global interrupt vector. For the 24-bit opcode parts there is a separate interrupt vector for each IF. In the case of the 12–16 opcode parts, the compiler generates code that will check each IF and when it finds both an IF and a corresponding IE set it jumps to the function defined by the programmer to handle that specific interrupt. Except for special cases discussed later, while the ISR is running, no other interrupts will be handled. There is another kind of global interrupt enable flag used only by the processor to prevent that. A specific assembly instruction is used to notify the processor that the ISR is done and it can return to where it was. This is how an ISR returns.

If all the IFs are checked and none are set with a corresponding IE set, then the compiler can be set up to call a default interrupt handler. The programmer uses **#int_default** to identify that function. This might happen if in the same moment an interrupt comes in the program disables the interrupt. Usually no one cares but if you do this is how it is handled.

There is also an **#int_global directive** to identify a global interrupt handler. When this is done it replaces the compiler-generated code to figure out what interrupt happened. When the **#int_global** is used then no other *#int_functions* can be defined. This is sometimes used if the programmer knows only a small number of interrupts can happen and thinks he or she can write tight code to do what is needed faster than the compiler code.

Handle Your Interrupts Right

It is important to understand how interrupts are generated, how the processor and compiler handle them, and what the impact is on your code. Some interrupts have special characteristics that need to be accommodated. For example, many of the PIC® processors have a general pin change interrupt. This is separate from the **int_ext** interrupts. As an example, the **int_rb** interrupt happens when any pin on port B changes. In this case there is only one IF flag to indicate some pin on the port changed. The programmer needs to figure out which pin caused the interrupt. To help, the compiler has the function **input_change_x()** that will return a bitmap of pins that changed since the last call. Some parts allow you to ignore changes on some of the port pins. In the compiler this is done with **enable_interrupts()** options. Another caution with the **int_rb** is that you must read the port inside the ISR otherwise the IF flag cannot be cleared. The **input_change_x()** does read the port.

The UART incoming **(int_rda)** and outgoing **(int_tbe)** interrupts also have an IF that cannot be cleared in firmware. If there is an incoming character in the buffer the only way to clear the interrupt is to read the character (like with a **getc()**).

If the special condition to clear an interrupt is not done the program will appear to hang because the ISR just keeps getting called.

One more consideration is a few interrupts also have a second enable that must be set in the peripheral itself. Usually this is so the programmer can select exactly what conditions that peripheral will interrupt on. An example of this is the comparator on many chips can be programmed to either toggle a pin, interrupt, or both when a difference in two analog voltages is detected.

Multiple Interrupt Considerations

Each family of chips deals with interrupts a little differently when it comes to multiple interrupts that happen at the same time. Here is a summary by family:

12-Bit Opcode Parts

Only a few support any interrupts at all. The few that do work exactly like the 14-bit parts.

14-Bit Opcode Parts

An interrupt can never interrupt an ISR on these parts. If two interrupts happen at the same time both will be serviced one at a time in a compiler-determined order. If the programmer wants to specify the order they are serviced, use the **#priority** directive.

16-Bit Opcode Parts

This family can operate in one of two modes. The default mode (called compatibility mode) is to work just like the 14-bit parts. The other mode is the dual priority interrupt mode. To use this mode the following directive is used:

```
#device  HIGH_INTS=TRUE
```

Now the chip uses two interrupt vectors identified as normal and high. The processor needs to know what interrupts are high and normal. This is done with an interrupt priority (IP) bit for each IF. In C any interrupts the programmer wants assigned to the high vector can be designated like this:

```
#int_ext  HIGH
```

Two interrupt handlers are generated by the compiler, one at each vector location. The new rule in this mode is simply that an interrupt designated as HIGH can interrupt a normal ISR. It is therefore possible to have two interrupt functions active at the same time. A HIGH cannot interrupt a HIGH but a HIGH can interrupt the main program or normal ISR. In total there could be three active threads in the program in this scenario.

HIGH interrupts are used for time-critical interrupts.

24-Bit Opcode Parts

As previously indicated, this family has a separate interrupt vector for each interrupt source. This simplifies the interrupt processing. On this chip an **#int_default** handler simply fills in all the interrupt vectors not specified by the programmer.

This family also supports a second, duplicate, set of interrupt vectors called the alternate vectors. A bit in a processor special function register (SFR) allows the programmer to select the active interrupt vector set. This is used when there are two separate programs in memory each with its own interrupts. In practice this is usually done with bootloaders. The bootloader when active will use its own interrupt handlers separate from the application. In C the *#build* directive is used to indicate which vector set to compile for.

Another feature of this family is that interrupt nesting can be enabled. With interrupt nesting enabled interrupts can interrupt other interrupts (but not themselves).

Yet another feature in this family allows an interrupt level (0–7) to be assigned to each interrupt handler. Instead of a hard interrupts-enabled or -disabled (GIE flag) there is available a service level number (0–7) that indicates which interrupts are enabled by level number. For example, if the service level is set to 3, only levels 3, 4, 5, 6, and 7 are serviced, the remaining act like the interrupt is disabled. This is used for complex programs where, depending on the current mode of the program, only certain groups of interrupts are enabled. For example, when performing a safety-critical operation maybe only safety-related interrupts are serviced. The level also determines which interrupt is invoked first if two interrupts happen at the same time. To set the level for an ISR, use:

```
#int_ext  level=3
```

Finally there is a group of interrupt sources defined in this family that is related to program errors. These are called traps but work like interrupts. For example, one of the traps is a divide by zero **(#int_matherr)**. If the code does a divide by zero a specific interrupt is generated. If the user did not define an interrupt handler for the trap then the chip is reset. There is no way to disable traps. They will always happen.

In C the compiler supplies a generic trap handler that can be used to alert the user to a problem. This can be helpful during debugging and the code can be modified for a production program. To use it simply do this:

```
#include <pcd_traps.c>
```

Latency

The time between the interrupt happening and the first line of code in your ISR being executed is called interrupt latency. This becomes important in time-critical applications. The major cause of interrupt latency is the code the compiler must insert at the interrupt vector before the C ISR can be called. The delays in the processor are insignificant compared to the handler code. Already discussed is that for many processors the source of interrupt must be determined. In addition to this the compiler saves a group of RAM locations and SFRs into a special save area. At the end of the ISR this data is restored. The SFRs saved are the ones that are routinely used to implement the C code. For example, the W register. The compiler uses a handful (2–8 bytes) of RAM locations as general scratch locations to save intermediate values during expression evaluation. These registers must be saved and restored so the code can continue right where it left off. The save and interrupt detect on a typical PIC16 program is around 30 instructions or at Fosc = 20 MHz, around 6 us. The program complexity affects how much is saved and with the number of defined interrupts this time could be double for you.

For applications that cannot tolerate a latency that large, there are some tricks that can be employed.

One option is to skip the compiler handling code entirely by using **#int_global**. In this case the programmer MUST be sure to save any SFRs it modified and to restore them. This will involve reviewing the assembly generated. The following is an example. It uses inline assembly that will be covered in Chapter 24. This should only be attempted when a programmer has a solid understanding of the assembly code and PIC® architecture. When you do this you cannot use any of the compiler-handled interrupts. You must handle all the interrupts in the program here.

```
#INT_GLOBAL
void isr()  {
   static int save_w;
   #locate save_w=0x7f  // location is not affected by bank switching
   static int save_status;
   #locate save_status=0x20

   #asm
   MOVWF save_w       //store current state of processor
   SWAPF status,W     //SWAPF writes to W Without affecting the status bits
   BCF   status,5
   BCF   status,6
   MOVWF save_status

           // Check to see if the timer 0 was the interrupt source
           // and if it was increment our counter
   if( interrupt_active(INT_TIMER0) )
      counter++;

   SWAPF save_status,W  // restore processor and return from interrupt
   MOVWF status
   SWAPF save_w,F
   SWAPF save_w,W
   #endasm
}
```

For the PIC18 parts there is another option. By enabling high priority interrupts you can define a single function at a single interrupt source and a high priority interrupt. In this case the compiler does only a minimal SFR save/restore and no check for the interrupt source. The user is still responsible for saving any RAM changed, but for a very simple ISR this may be nothing. The FAST option is used to designate such a handler. Here is a simple example:

```
#int_ext  fast
void  int_isr(void) {
  output_toggle(PIN_C3);
  got_it=TRUE;
}
```

Reentrancy

Assume you have a function named **foo()** and you have an interrupt handler named **isr()**. Further assume both **main()** and **isr()** call **foo()** . The problem scenario is when **main()** calls **foo()** and execution is inside **foo()** when the interrupt comes in.

At this point **isr()** calls **foo()** and we have reentrancy: a case where two active threads are executing in the same function. This is a problem on the PIC® because the stack is not used for the local data so the second call will disrupt data from the first call.

In order to prevent this situation, the compiler will disable interrupts before the **main()** call to **foo()** and enable them afterwards. This makes the call safe but can introduce an additional interrupt latency in the case the interrupt comes in while executing **foo()**. To help the programmer to identify these situations the compiler generates a warning message when this is done.

Be aware that the functions identified in these warnings are sometimes compiler-generated functions such as multiply. Compiler-generated functions always start with a **@**.

Compatibility Notes

The closest thing standard C has to dealing with interrupts is a library in **signal.h**. In reality it is not practical for microcontroller use. Most microcontroller C compilers do not even make an attempt to use **signal.h** for interrupt handling.

Some compilers ignore the issue, forcing programmers to write code directly accessing the SFRs to set up interrupts and putting code at the interrupt vector to handle the interrupts. Compilers with interrupt support all do it differently. Don't expect any compatibility when it comes to interrupts.

Summary

- Interrupts pause program execution in response to some hardware event and cause an interrupt service routine (function) to execute.
- Interrupts happen independently (asynchronously) from program execution.
- Interrupts return to the point they interrupted when the ISR completes.
- Programmers must define an interrupt handler for each interrupt source, enable the interrupt source, and enable global interrupts.
- The time spent in an ISR is added to the time the main program takes to execute.
- Noise generating multiple interrupts may be accounted for with interrupt handlers.
- Extreme care must be taken when multi-byte data is accessed inside and outside an ISR.
- Interrupt handling is controlled by an interrupt flag, an interrupt enable bit, and the global interrupt enable bit.
- Interrupt handlers are invoked from the interrupt vector. This vector is the link from the hardware interrupt to the firmware handler.

- Some PIC®s have additional interrupt features such as a dual priority interrupt, interrupt levels, and interrupt nesting.
- Interrupt latency is the time from the hardware signal until the ISR code begins execution.

Exercise 17-1

Objective: Write programs that use interrupts to perform some basic tasks.
Requires: E3 module, USB cable, PC, frequency generator (if available).

Steps/Technical Procedure	Notes
1. Write a program that will output a message every 3 s indicating the number of times each of two of the push buttons was pressed and the number of times they were released. A total of four numbers should be output. On this processor the INT1 and INT2 pins are C1 and C2. Both are connected to push buttons. **HINTS:** • INT1 is the INT_EXT1 interrupt in the compiler. • The EXT interrupts happen only on one edge of a pulse. To get the other edge you will need to change the detect edge every time an interrupt comes in. • For this program, since we have not dealt with timers yet use a delay_ms(50) inside the ISR to debounce.	
2. The program is to have two float global variables X and Y. A push-button ISR should be set up to execute each time the button is pressed. A local static variable is to keep degrees, starting at 0. Each button push increments degrees and in the ISR a new X,Y as a position on a circle with a radius of 5000. The main program uses X and Y to calculate the angle in degrees. If the integer value of the angle (0–359) changes from the last value displayed then the main program should output the new angle to the screen. **Technical Information:** sin() returns the Y position for a given angle in radians on a 1-unit circle. cos() returns the X position. arcsin() and arccos() are the inverse functions. There are 2 * 3.1415 radians in a circle. There are 360 degrees in a circle. **HINTS:** • The key to this exercise is to make sure the interrupt function does not disrupt the partial calculations in the main program.	

3. Write a program that will measure the frequency of pulses coming into the C0 pin of the E3 board. Interrupts must be used. Make sure the frequency generator is set to generate a signal from 0 to 5 V. DO NOT GENERATE A SIGNAL BELOW GROUND. **HINTS:** • The program should use interrupts to count pulses and the main program will each second display the count and clear the counter. • If a frequency source is not available press the button as fast as you can to test.	

Quiz

(1) Of the following, which are very good applications for interrupts?

- (a) Button presses on a blender to select the speed
- (b) Button press for an emergency stop
- (c) Buttons on a clock to set the hours and minutes
- (d) The fill button on a beverage machine
- (e) All of the above

(2) Which of the following statements are true?

- (a) Only single-byte variables can be shared between interrupt functions and main()
- (b) No variables should be shared between interrupt functions and main()
- (c) No variables may be shared between different interrupt functions
- (d) No special care need be taken with variables shared between different interrupt functions
- (e) Interrupt functions cannot access global variables

(3) A program designed to count button presses by way of an interrupt seems to always show one too many presses. What is the likely solution?

- (a) A `clear_interrupt()` should be done before the interrupt is enabled
- (b) The global interrupt should be enabled after (not before) the specific interrupt level
- (c) The count variable should be initialized to -1 not 0
- (d) A `clear_interrupt()` should be done at the top of the ISR
- (e) A debounce algorithm needs to be employed

(4) On a PIC16 class processor how many IF (interrupt flag) bits can be set at the same time?

- (a) One
- (b) Two

(c) One for each IE flag set
(d) One for each interrupt source
(e) One for each ISR actively executing

(5) A program has a single interrupt, a timer that interrupts 50 times per second. The ISR including latency takes 10 ms to execute. What percentage of the total CPU time is lost due to interrupts?
(a) Less than 1%
(b) 1%
(c) 10%
(d) 20%
(e) 50%

(6) A program has a single interrupt, a timer that interrupts 50 times per second. The ISR including latency takes 10 ms to execute. How long will a **`delay_ms(110)`** take to execute?
(a) 110 ms
(b) 160 ms
(c) 220 ms
(d) Between 195 and 245 ms
(e) Between 160 and 170 ms

(7) An ISR is triggered by B0 going low. The first thing the ISR does is to set pin B5 high. It then does a math calculation that takes 20 us. Finally B5 is set low before the ISR ends. The measured ISR latency is 5 us. Looking at B0 and B5 on a scope, what will you see?
(a) B0 goes low and at the same time B5 goes high for 20 us and then low
(b) B0 goes low and at the same time B5 goes high for 25 us and then low
(c) B0 goes low and at the same time B5 goes high for 25 us and then low
(d) B0 goes low and 25us later B5 goes high for 5 us and then low
(e) B0 goes low and 5 us later B5 goes high for 20 us and then low

(8) Under what circumstances would a programmer disable interrupts?
(a) To protect manipulation of variables shared with an ISR
(b) When the interrupt is no longer needed by the program
(c) During a critical time when the most CPU time is required
(d) All of the above
(e) There is never a good reason to disable interrupts

(9) On a PIC18 processor with dual priority interrupts the INT pin is connected to a button and the INT1 pin is connected to pin B5. INT1 is set up as a high priority interrupt and INT is normal. Both trigger on the falling edge. The main program sets B5 high, clears interrupts, and enables interrupts. It then sits in a loop. The following is the ISR code. Once the button is pressed, how long will the B3, B4, and B5 pulses be low?

```
#INT_EXT1 HIGH
void high_isr(void) {
    output_low(PIN_B4);
    delay_ms(5);
    output_high(PIN_B4);
    output_high(PIN_B5);
}
#INT_EXT
void high_isr(void) {
    output_low(PIN_B3);
    delay_ms(10);
    output_low(PIN_B5);
    delay_ms(5);
    output_high(PIN_B3);
}
```

(a) 20 ms, 5 ms, 5 ms
(b) B3 remains low forever, 5 ms, 10 ms
(c) 15 ms, 5 ms, 10 ms
(d) 15 ms, 5 ms, 5 ms
(e) B3 remains low forever, 5 ms, 5 ms

CHAPTER 18

Timers/Counters

Timers and *counters* are a basic peripheral unit on microprocessors. We are essentially talking about a counter, and when the source is a clock it is called a timer. Technically when the source is not a clock it is a counter. The Microchip data sheets always call it a timer so we will do so for the rest of this chapter.

The essential characteristic of the timer is the number of bits. For example, an 8-bit timer can go from 0 to 255. Most processors have several timers, each with slightly different capabilities. Most processors also have additional peripheral units directly connected to the timers for advanced functionality. Those will be discussed in Chapter 19.

The diagram in Figure 18.1 is of a fully loaded timer. Not all timers will have all these features; however, we will use this diagram to walk through the key timer components.

Timer Components

The Counter Core

At the center we have the counter itself. Simply put, each pulse from the left causes the counter to increment by one. At any time the firmware can read the timer value or reset it to any number desired. All the PIC® processors count up. Some other processors will count down.

The Counter Period

The PERIOD register indicates the reset value for the counter. When the counter reaches this number it resets back to 0. Most timers do not have the period register, so the reset value is the largest counter number plus one. For example, on an 8-bit counter the values will go from 0 up to 255 and then the next pulse flips the counter back to 0. This flip is called an *overflow*. If the device had a period register and it was set to 200 then the count would go from 0 to 200 and then back to 0 on the next pulse.

The Post-scaler

When the counter restarts it generates a signal out. This is represented to the right on the diagram. Normally this goes directly to an interrupt flag so an interrupt is generated if enabled.

Embedded C Programming. http://dx.doi.org/10.1016/B978-0-12-801314-4.00018-1

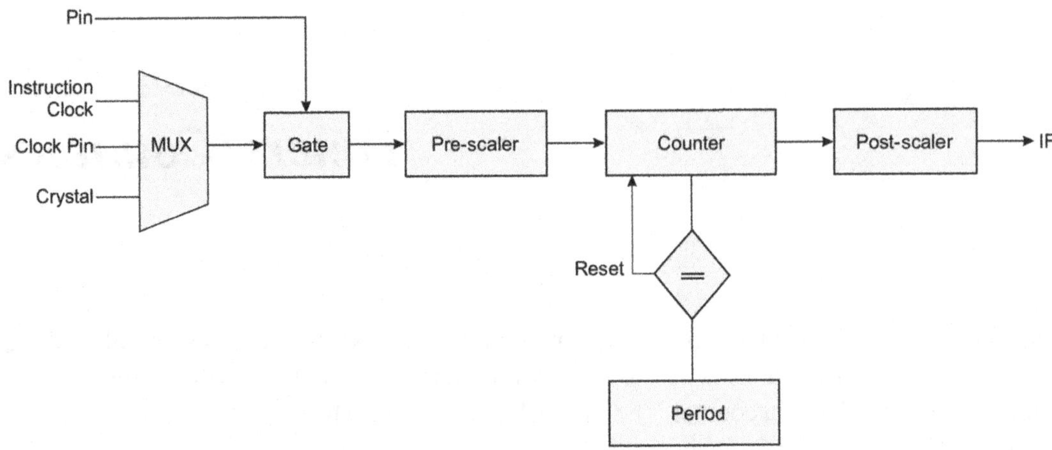

Figure 18.1: Fully loaded timer diagram.
MUX = multiplexer.

Some timers have a post-scaler as shown in the diagram. If the post-scaler value is 4 then that means for every four timer resets there will be one setting of the interrupt flag (IF). This slows down how many interrupts happen. For example, say we have an 8-bit timer where there is one pulse in every 1 us. Then there is a signal out every 256 us. With a post-scaler of 4 this means an interrupt every 1024 us, or about once a millisecond. The post-scaler allows for larger gaps between interrupts without adding bits to the counter.

The Pre-scaler

On the left side of the counter we find many units have a pre-scaler. It works like the post-scaler in that only for some programmed number of incoming pulses does a pulse get passed to the counter. If the pre-scaler is set to 16, for example, and the source is a 1-us clock, the counter increments every 16 us, and if 8 bit, it overflows every 4096 us. The pre-scaler can be very useful in getting a clock in the range you need. You can see depending on what you want to do with the timer you will want to obtain an ideal range and accuracy. This is aided by a good selection for the pre- and post-scalers.

The Gate

To the left of the pre-scaler we have a gate. The timers that have a gate will have the gate connected to an external pin. That pin controls whether pulses are seen by the counter at all. Imagine we have the 1-us clock as a source with no pre-scaler. Say a pulse comes in on the gate pin for 50 us. That opens the gate for 50 us and the counter sees around 50 pulses from the clock before the gate closes. The counter value will then be 50, the size of the pulse. There are other uses for the gate and better ways to measure a pulse (see Chapter 19) but this example illustrates how the gate works.

The Multiplexer

Before the gate we have a multiplexer (MUX). This selects the source of signal. The basic sources you will find on the PIC® device are as follows:

- The PIC® instruction clock. On most PIC® MCUs this is Fosc/4. For example, a 20-MHz PIC16 will have a 5-MHz instruction clock, or a 200-ns pulse rate to the pre-scaler.
- External pin. Usually the rising edge is what causes the timer increment. Check the electrical specifications in the data sheet for minimum low and high times and maximum frequency. Many parts offer the option of synchronizing the counter increments with the instruction clock.
- External crystal. Why an external crystal when you have a crystal already for your processor? The answer is sometimes it is worth it to get a clock at just the right frequency. For example, say you are doing an alarm clock. With a 20-MHz crystal into a 16-bit counter and a /64 pre-scaler, you get an interrupt every 0.8388608 seconds. Using a 32.768-kHz crystal (yes they make them, just for this) and a /2 pre-scaler, you get an interrupt every 1.0000000 seconds. Very nice if you want to keep accurate time.

PIC® Specifics

Table 18.1 shows the typical timers found in a typical chip in each family group. Some chips in each group have several more timers.

C Code

Let's start with a simple C program so we can see what the function calls look like:

```
#include <E3.h>

void main(void) {
   int16 time;
   int32 y;
   int32 x = 12345678;

   setup_timer_1( T1_INTERNAL | T1_DIV_BY_1 );
   set_timer1(0);
   y = x*3;
   time = get_timer1();
   cout << "The multiply took " << time << " ticks." << endl;
}
```

The setup call sets the source to the internal instruction clock and the pre-scaler to 1. The or bar, |, is used to combine some number of mode options on many built-in function calls.

Table 18.1 Typical timers found on a PIC® MCU.

Name	Bits	Pre-scaler	Period	Post-scaler	Gate
12-Bit Opcode					
0 aka RTCC	8	1, 2, 4, 8, 16, 32, 64, 128, 256			
14-Bit Opcode					
0	8	1, 2, 4, 8, 16, 32, 64, 128, 256			
1	16	1, 2, 4, 8			Yes
2	8	1, 4, 16	Yes	1–16	
16-Bit Opcode					
0	16	1, 2, 4, 8, 16, 32, 64, 128, 256			
1	16	1, 2, 4, 8			
2	8	1, 4, 16	Yes	1–16	
3	16	1, 2, 4, 8			
24-Bit Opcode					
1	16	1, 8, 64, 256	Yes		Yes
2/3	32	1, 8, 64, 256	Yes		Yes
4/5	32	1, 8, 64, 256	Yes		Yes
PIC24 x/y timers may be configured as one 32-bit timer or two 16 timers.					

For this chip and timer 1, the following options are from the device header file:

MUX options:

```
T1_DISABLED
T1_INTERNAL
T1_EXTERNAL
T1_EXTERNAL_SYNC
```

To set up for external crystal:

```
T1_CLK_OUT
```

Pre-scaler options:

```
T1_DIV_BY_1
T1_DIV_BY_2
T1_DIV_BY_4
T1_DIV_BY_8
```

The **set_timer1()** call sets the timer value to 0.

The math looks like this:

Timer tick = 1/(Fosc/cycles_per_instruction) ∗ prescale
Overflow time = tick ∗ 2^{BITS}
Interrupt time = overflow_time ∗ postscale
Interrupt rate = 1/overflow_time

The E3 has Fosc = 48 MHz and, like most chips, the E3 chip has four cycles_per_instruction and the pre-scale was set to 1. BITS is 16 and there is no post-scale (1) so the math works out to:

Timer tick = 83.33 ns
Overflow time = 5.461 ms
Interrupt rate = 183.1 Hz

The **get_timer1()** reads the timer value. There will be some error in this time due to overhead in the function calls. The following code fixes that error and outputs the time in us.

```
#include <E3.h>

void main(void) {
   int32 time;
   int16 overhead;
   int32 y;
   int32 x = 12345678;

   setup_timer_1( T1_INTERNAL | T1_DIV_BY_1 );
   set_timer1(0);
   overhead = get_timer1();
   set_timer1(0);
   y = x*3;
   time = get_timer1();
   time = time - overhead;
   time = (time+6)/12;                 // convert to us assumes Fosc=48mhz
   cout << "The multiply took " << time << " us." << endl;
}
```

Delay Using Timer

We have been using **delay_us()** a lot; however, in all cases we are not able to do anything during the delay. The following code sample performs a **delay_us(100)**; however, breaks out of the delay if pin C0 goes low:

```
time = 100 * 12;           // 100us times ticks per us (Fosc=48mhz)
set_timer1(0);
while( (get_timer1()<time) && (input(PIN_C0)) ) ;
```

Precision Loop

It is a common design method to have a main program loop that executes at an exact frequency. For example, 50 times a second (20 ms). This helps to control the timing of various tasks that need to be done. The following code uses the timer to control the loop frequency:

```
setup_timer_1( T1_INTERNAL | T1_DIV_BY_8 );    // at Fosc=48mhz each tick
while(TRUE) {                                  // is 0.667 us
   set_timer1(0);
   perform_tasks();
   while( get_timer1() < 30000) );             // Wait for 20ms mark
}
```

Interrupts

Timers can generate a single interrupt when they overflow. Some timers have a post-scaler that will divide down the interrupt frequency. The following example shows using the timer interrupt to extend the number of bits in the timer. We use a 16-bit timer, but then by keeping a count of overflows it is like a 32-bit timer.

```
int16 timer1_high =0;

#int_timer1
void timer1_isr(void) {
   timer1_high++;
}
void main(void) {
   int32 time;

   setup_timer_1( T1_INTERNAL | T1_DIV_BY_1 );
   enable_interrupts(INT_TIMER1);
   enable_interrupts(GLOBAL);
   while(TRUE) {
      while(input(PIN_C0));                    // Wait for button press
      time = ((int32)timer1_high<<16) | get_timer1();
      cout << "Timer value: " << time << endl;
   }
}
```

Since there are limited timers available in the chip, a common practice is to have a high resolution long range timer like the above free-running. Instead of resetting the timer value, to time things you just copy the start value and subtract it from the end value. This way many different parts of the program can just use the same timer.

This code has a problem however. Consider the case where the **timer1_high** is 0x0005 and timer 1 is 0xFFFE. As the expression is evaluated the 0x0005 is shifted up to give us 0x00050000 and then just before we do the | timer overflows, **timer1_high** goes to 6 and the timer wraps around to 0. The interrupt service routine (ISR) returns and now we finish the expression. The **get_timer1()** gives us a 2 and the time variable is 0x00050002. The right value should be between 0x0005FFFE and 0x00060002. Our value is way off. Chapter 17 warned us about problems like this.

The following function will solve the problem:

```
int32 get_timer1_32(void) {
   int16 high;
   int16 low;

   do {
      high=timer1_high;
      low=get_timer1();
   } while (high!=timer1_high);
   return (int32)high<<16 | low;
}
```

Study it to figure out why this code works.

Interrupts at Specific Rates

It is very common to want an ISR to execute at a fixed rate. This is easy if the rate happens to be equal to the overflow rate. For other rates we need to do some timer manipulation. Here is an example of an ISR set up to execute every 10 ms and toggle a pin.

```
#int_timer1
void timer1_isr(void) {
   set_timer1( 50536+get_timer1() );
   output_toggle(PIN_B4);
}

void main(void) {
   setup_timer_1( T1_INTERNAL | T1_DIV_BY_8 );   // Fosc=48mhz
   enable_interrupts(INT_TIMER1);
   enable_interrupts(GLOBAL);
   while(TRUE) ;
}
```

Let's start by looking at the math to get the 50536. It is:

$$reload\_value = 2^{BITS} - isr\_rate/(Fosc/cycles\_per\_instruction)$$
$$reload\_value = 65536 - 0.01 * (48000000/4/8)$$
$$reload\_value = 65536 - 15000$$
$$reload\_value = 50536$$

It takes 15000 ticks to pass 10ms and we want to set the timer up so it overflows in 10ms. This means we subtract 15000 from the overflow value.

In the code we then add the timer 1 value. The current value of timer 1 represents the interrupt latency. The interrupt happened when timer 1 was 0. Whatever value it is now is the latency. By adjusting the next interrupt by the latency, we will get a much more accurate rate.

This is easier if the timer has a programmable period. Here is an example of a 100-us rate timer using timer 2. Timer 2 is only 8bit, so to get 10 ms we would need a counter in the ISR

```
#int_timer2
void timer2_isr(void) {
   output_toggle(PIN_B4);
}
void main(void) {
   setup_timer_2( T2_INTERNAL | T2_DIV_BY_1,    // Fosc=48mhz
                  150,    // Period value
                  8 );    // Postscaler
   enable_interrupts(INT_TIMER2);
   enable_interrupts(GLOBAL);
   while(TRUE) ;
}
```

The math here is:

rate = prescale ∗ postscale ∗ period / (Fosc/cycles_per_instruction)
rate = 0.0001 seconds

Interrupt at a Specific Time

Using the above reload technique, you can now see how easy it is to set up a timer to interrupt at a specific time. One use of this is when controlling a relay that has an AC load. To reduce contact wear and EMI it is best to switch the relay when the AC power crosses the zero point. With 60Hz AC, this happens 120 times a second or every 8.333ms. The circuit provides a digital signal at the zero-crossing. Relays can take some time to engage so activating the relay when the interrupt happens will cause it to make contact too late. The best way to handle it is to catch the next zero-cross minus the engage time.

Here is our example with B5 controlling the relay and B0 supplying the zero-cross detect. We use B1 to connect to a button that goes low when pressed. The relay will track the button in the following example. Use Figure 18.2 to match the code up to the timing.

```
#define TICKS_PER_US  6
#define ZERO_CROSS_TIME     8333  //us
#define RELAY_ENGAGE_TIME   3200  //us

#int_timer1
void timer1_isr(void) {
   output_bit(PIN_B5,input(PIN_B1));
   disable_interrupts(INT_TIMER1);
}

#int_ext
void ext_isr(void) {
   set_timer1( 0x10000 -
          ((ZERO_CROSS_TIME-RELAY_ENGAGE_TIME)/TICKS_PER_US) );
   clear_interrupt(INT_TIMER1);
   enable_interrupts(INT_TIMER1);
   disable_interrupts(INT_EXT);
}

void main(void) {
   output_low(PIN_B5);
   setup_timer_1( T1_INTERNAL | T1_DIV_BY_4 );  // Fosc=48mhz  Tick=167ns
   enable_interrupts(GLOBAL);
   while(TRUE)  {
       while(!input(PIN_B1)) ;
       clear_interrupt(INT_EXT);
       enable_interrupts(INT_EXT);
       while(input(PIN_B1)) ;
       clear_interrupt(INT_EXT);
       enable_interrupts(INT_EXT);
   }
}
```

Virtual Timers

The compiler has a built-in library that can be set up to manage a timer and automatically deal with extending the timer range using an interrupt. We call the new timer a *virtual timer*. Although it uses a physical timer to do the work there is a virtual timer with a user-specified range and resolution. Here is an example of a virtual timer with a tick value of 1 ms:

```
#use timer(timer=1, tick=1ms, bits=32, ISR)
```

This provides for a timer good to 49 days. The function **get_ticks()** returns the virtual timer value.

Here is a interrupt function that rejects interrupts that happen within 100 ms of each other. This is a good technique for filtering signals or debouncing buttons.

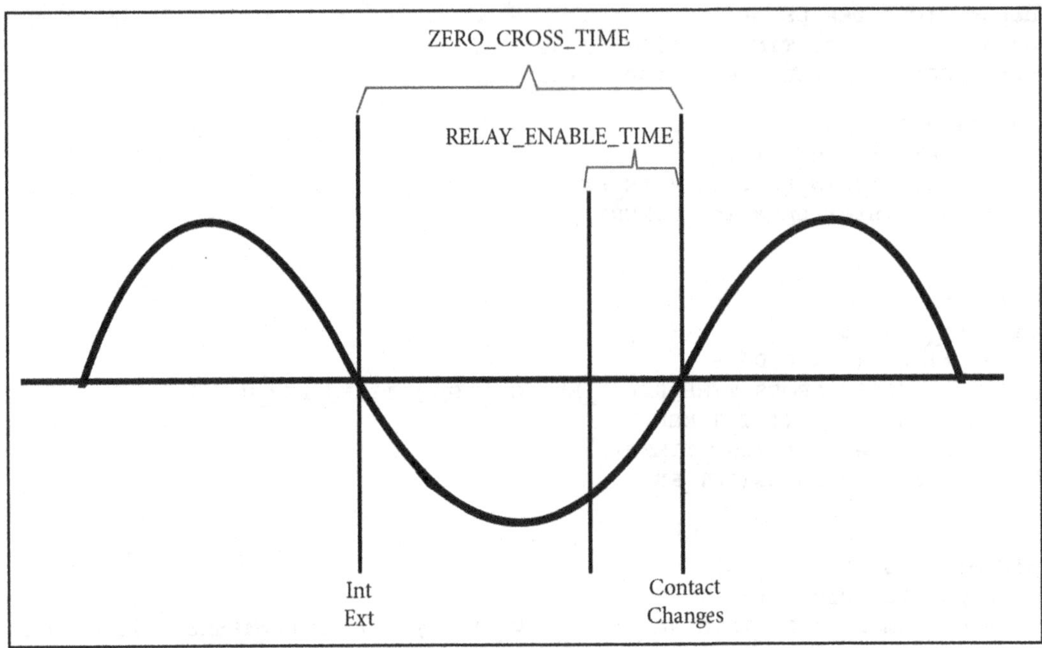

Figure 18.2: Timing diagram for switching on the AC zero-crossing.

```
#use timer(timer=1, tick=1ms, bits=32, ISR)
#int_ext
void ext_isr(void) {
   static int32 last_time;
   if( abs(get_ticks()-last_time)<100 )
      return;
   last_time = get_ticks();
   output_toggle(PIN_B5);
}

void main(void) {
   enable_interrupts(INT_EXT);
   enable_interrupts(GLOBAL);
   while(TRUE);
}
```

Summary

- Counters count input pulses.
- Timers are counters with a clock as the input source.
- Counter/timers restart at 0 when the period is reached. Without a period register this is 2^n where n is the number of bits in the counter.

- Upon a reset to 0 an interrupt flag can be set.
- With a post-scaler the interrupt flag setting may be delayed by a programmed divisor.
- Post-scalers can be used to divide the input pulses to the counter.
- Gates are sometimes used to disable inputs to the pre-scaler.
- In C, functions like `setup_timer_x()`, `set_timerx()`, and `get_timerx()` are used to manage the timer.
- Virtual timers can be set up with `#use timer` and used with `get_ticks()`.

Exercise 18-1

Objective: Learn how to use a basic timer from C.
Requires: E3 module, USB cable, PC.

Steps/Technical Procedure	Notes
1. Write a program that will start by asking the hours, minutes, and seconds for the current time. Then each time the button is pressed output a message with the current time in hours, minutes, and seconds. This program should use the E3 48-MHz clock to maintain the time. A second crystal for the clock is not permitted. **HINTS:** • Keep the time as a large variable with the number of seconds since midnight. Do the conversion to/from hours, minutes, and seconds as needed.	
2. Write a program that will first light the yellow LED and wait for a button press. It will then wait a random time from 5 to 15 seconds before the green LED it lit. Once lit the user should press the button again. The program will then display at the PC the reaction time in milliseconds from LED on to button press. A good method to find a random number is to start a timer on power-up and to grab the timer value when a button is pressed. The desired number of bits can then be masked off and scaled. For example, the number might represent tenths of a second. **HINTS:** • To get the required range the timer will need to be extended to 32 bits via an ISR. • Timer 2 has a period register that can be used to limit the range of the timer. This can help a bit when it is used for a random number generator within some range.	

Quiz

(1) Consider a PIC16 processor with Fosc = 20 MHz, an 8-bit timer with a pre-scaler of 4, and no post-scaler. Around how many interrupts will happen per second?
  (a) 1221
  (b) 1953
  (c) 4883
  (d) 78,125
  (e) 625,000

(2) Assume a PIC16 class processor running at Fosc = 4 MHz using 16-bit timer 1 with pre-scaler options of 1, 2, 4, and 8. Without using interrupts or the interrupt flag, what is the longest event that can be timed with the timer?
  (a) 1 us
  (b) 64 us
  (c) 512 us
  (d) 65.536 ms
  (e) 524.288 ms

(3) What design trade-off is made by choosing to use a larger pre-scaler?
  (a) Reduced timer range
  (b) Reduced precision
  (c) Fewer post-scaler options
  (d) Increased interrupt latency
  (e) Fewer bits in the counter

(4) Given the following code, with all interrupts disabled what will be the pulse width on B5 for a PIC16 class part running at Fosc = 4 MHz?

```
setup_timer_1( T1_INTERNAL | T1_DIV_BY_1 );
set_timer1(0);
output_high(PIN_B5);
clear_interrupt(INT_TIMER1);
while( !active_interrupt(INT_TIMER1) || (get_timer1()<32768) );
output_low(PIN_B5);
```

  (a) Less than 20 us
  (b) 32.7 ms
  (c) 65.5 ms
  (d) 98.3 ms
  (e) B5 never goes low

(5) The following is an alternate function to read an extended timer. What is the flaw in this code?

```
int32 get_timer1_32(void) {
        int32 result;
        disable_interrupts(GLOBAL);
        result=(int32)timer1_high<<16 | get_timer1();
        enable_interrupts(GLOBAL);
        return result;
}
```

(a) There is no way to ensure what order **timer1_high** and **get_timer1()** are read
(b) Disabling the interrupts will increase the ISR latency
(c) The timer value may change between the assignment to result and the return of result
(d) The **disable_interrupts()** does not stop the timer
(e) There is no flaw in this logic

(6) A vehicle has wheels with a 5-foot circumference that have a sensor that pulses once each revolution. This signal is fed into a counter input on a PIC® with no pre-scaler and Fosc = 16 MHz. Every second the firmware reads and clears the counter. What is the formula for determining the vehicle speed in MPH?
(a) MPH = (5281/get_timer1() * 5) * 60 * 60
(b) MPH = 60 * 60 * 5 * get_timer1()/5281
(c) MPH = (60 * 60 * 5 * get_timer1()/4)/5281
(d) MPH = (5281/get_timer1()) * 60 * 60
(e) MPH = (5281/get_timer1()/4) * 5 * 4000000

(7) What two timer concepts cannot be used at the same time?
(a) Pre-scaler and post-scaler
(b) Gate and pre-scaler
(c) Post-scaler and period register
(d) Post-scaler and extended bits
(e) Gate and instruction clock

(8) Assume a PIC16 class processor running at Fosc = 4 MHz using 16-bit timer 1 with pre-scaler of 1. This timer is used to measure the width of a pulse. What is the best accuracy in the measurement that can be expected?
(a) +/−250 ns
(b) +/−1 us
(c) +/−256 us
(d) +/−65,536 us
(e) The timer measurement will be exact

(9) Assume a PIC16 class processor running at Fosc = 4 MHz using 8-bit timer 2 with a pre-scaler of 2, period of 50, and post-scaler of 5. How many interrupts per second will be generated?

(a) 5
(b) 500
(c) 1000
(d) 2000
(e) 2560

(10) A program is using timer 1 free-running (never set by firmware to a value) to time a pulse. The following is the code used. What modification needs to be made to account for the situation where the timer overflows between the start and end reading?

```
while(!input(PIN_B1));
start = get_timer1();
while(input(PIN_B1));
end = get_timer1();
time = end-start;
```

(a) time = (end<start) ? start-end : end-start
(b) time = abs((signed int16)(end-start))
(c) time = end+0 × 10000*(end<start)-start
(d) Any of the above will work
(e) The original code will work as is

CHAPTER 19

Advanced Timing

Most processors have some additional peripherals attached to a timer. That is because with a little extra hardware help the timer can do some very powerful things. This chapter will cover three peripherals that are a part of most of the PIC® processors.

- *PWM*—A pulse width modulator is a commonly used signal in hardware. It is in effect a square wave generator.
- *Capture*—This feature allows the hardware to be programmed to copy the timer value to a holding register when a hardware event, such as a pin changing state, happens.
- *Compare*—This feature will cause some hardware event, like a pin changing state, to happen when the timer reaches a certain value.

The PIC16/18 class parts combine all three of these features into one unit called the CCP (capture/compare/PWM). Some parts have an additional unit for just PWM where the PWM features are more sophisticated than a normal CCP unit has. Some data sheets will call the CCP unit ECCP (E for enhanced) when the PWM part has some of the advanced features.

The PIC24 class parts have separate *input capture* (IC) and *output compare* (OC) modules. The latter can also be used for PWM. Like the PIC16/18 parts, some PIC24 class parts have a separate feature-rich PWM.

PWM

We have already produced a square wave using **`output_high()`**, **`output_low()`**, and a **`delay_ms()`**. The advantage of a hardware PWM is the pulses continue while the program is doing other work. There are two characteristics of a PWM signal, the frequency and the duty. The duty is usually expressed as a percentage of the time the pulse is high. Figure 19.1 is an illustration of various duties with the same frequency.

In hardware it is an easy task to convert a PWM signal to a voltage where the voltage is proportional to the duty. This gives us the ability to generate a specific analog voltage from the digital signal. Figure 19.2 is an example circuit.

Another use for the PWM is motor control. A multi-phase motor energizes different coils in sequence to turn the motor. A PWM signal can be used to generate the required sequence for

Embedded C Programming. http://dx.doi.org/10.1016/B978-0-12-801314-4.00019-3

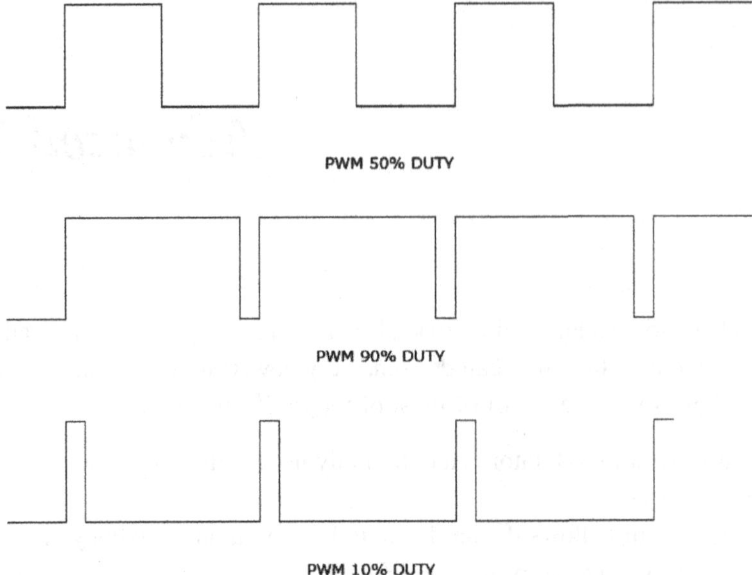

Figure 19.1: Pulse width modulation scope traces.

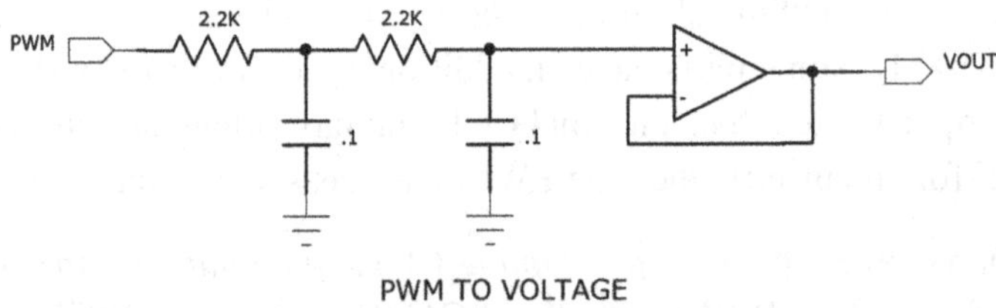

Figure 19.2: PWM to voltage schematic.

the motor. To save hardware, some PIC® devices have the ability to output two to four PWM output signals that are derived from the single PWM. For example, one pin might be high while another is low and vice versa. When this is done sometimes you will specify a dead time where when the signals switch both are inactive for a short period. Parts with an ECCP have this capability and other parts have a specific PWM-only module meant for motor control. Motor control is beyond the scope of this book so we will not be going into any further details on that here. It is an easy task to check the device header file for the part you will be using to find the built-in functions for motor control.

For a slow PWM, typically a timer interrupt is set up and the interrupt service routine (ISR) toggles a pin. For faster frequency PWMs the built-in hardware is used.

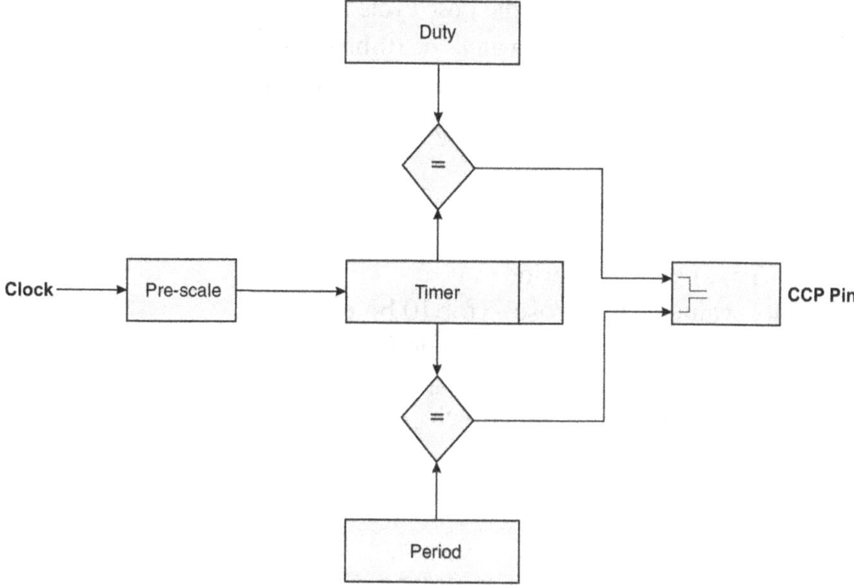

Figure 19.3: PIC MCU CCP module in a PWM configuration.

Figure 19.3 is a diagram of the PIC16/18 CCP peripheral in the PWM mode.

The clock and pre-scaler work as they usually do for a timer. When the period register equals the timer value in addition to the normal timer operation (interrupt flag (IF) and reset) it will set the CCP pin high. This is the start and end of a full cycle. The period register therefore determines the PWM frequency.

A comparison is also made from the timer to the duty register, and when equal the CCP pin is pulled low. This happens in the middle of a period and controls how long the signal is high during the period. It is clear to see how this simple addition of a little hardware creates a very nice PWM capability. Note that the hardware duty register is a timer tick value, not the usual percentage a duty is thought of as.

There is one additional feature in the CCP peripheral. The duty register is 10 bits in order to get a higher resolution for the duty. The resolution is derived from how many steps there are from 0% to 100% in the duty setting. The timers used by the CCP are always 8 bit (usually timer 2). For just this function (PWM) the hardware comes up with two more bits that can be used in the duty comparison. If a pre-scaler is used the bits come from the pre-scaler counter. For example, if you have a divide by 4 then there is a 2-bit counter in the pre-scaler that can be used for a higher resolution to the counter. If the timer value is 50 and the pre-scaler counter is 2 then this is like 50.5. If there is no pre-scaler then the instruction clock cycle counter is used. For every four Fosc cycles one instruction is executed and one pulse is sent to the timer. The processor keeps track of the Fosc cycle it is on so another 2 bits are available from there. The hardware can operate in an 8- or 10-bit duty mode depending on the program needs. In C we base the mode on the type of the argument to the duty function (int8 or int16).

Here is some C code that sets up a 10-kHz PWM with a 50% duty.

The key formulas are:

cycle_time = (4/Fosc) *t2div*(period+1)
high_time = duty_value*(t2div/Fosc) (for 10 bit duty)
high_time = duty_value*4*(t2div/Fosc) (for 8 bit duty)
Frequency = 1/cycle_time
Duty (as a percent) = 100*(high_time/cycle_time)

The code is:

```
setup_ccp1(CCP_PWM);    // Configure CCP1 as a PWM
setup_timer_2(T2_DIV_BY_4, 249, 1);
          //    Fosc=40000000, t2div is 4 and period=249
          //     (4/40000000)*4*250 = 100 us or 10 khz
set_pwm1_duty((int16)500);
          //  100*(500*(4/40000000)/0.0001) = 50%
```

To change the duty to 25%, just do this:

```
set_pwm1_duty((int16)250);
```

On the 24-bit opcode parts it works a little differently. They do not have a CCP peripheral but rather separate capture and compare peripherals. The compare peripherals have a mode to do PWM. Both the timer and the duty register are 16 bits. The key formulas are:

```
cycle_time = (cycles_per_instruction/Fosc)*txdiv*(period+1)
high_time = (cycles_per_instruction/Fosc)*txdiv*duty_value
Frequency = 1/cycle_time
duty (as a percent) = 100*(high_time/cycle_time)
```

The code for a PIC24 with Fosc = 40 MHz is:

```
setup_timer2(TMR_INTERNAL | TMR_DIV_BY_1, 999); // 10khz
setup_compare(1, COMPARE_PWM | COMPARE_TIMER2);
set_pwm_duty(1,500);  // 50%
```

Using the PWM Library

The above is good to help understand how the chip works and is helpful when you need to perform special functions. If all you need is a standard PWM, however, the compiler can generate code for your specific needs. The following is code to allow the compiler to set up the hardware for you, saving on some of the math:

```
#use pwm( timer=2, ccp=1, frequency=10khz )
```

The duty is set like this:

```
pwm_set_duty_percent( 500 );
```

In this case the 500 is 50.0%. The argument is not the high time as above but rather tenths of a percent.

The frequency can be changed with the following function:

```
pwm_set_duty_frequency( 5000 );   /// 5khz
```

Capture

The capture peripheral again is simple in concept but can provide some powerful capabilities. It works by simply making a copy of the timer value to a holding register when a pin goes from low to high or high to low. Figure 19.4 is the block diagram.

The timer is always a 16-bit timer, as is the holding register. On most chips you can specify a pre-scaler of 1, 4, or 16. This means you can capture the timer value on, for example, the fourth change of the CCP pin from low to high. At the same time the timer value is copied to the holding register an interrupt can be generated.

Usually you will want to use the interrupt and grab the value from the holding register. This is because the capture peripheral remains armed and if another event happens a new value is copied to the holding register.

This is used for very accurate timing of a hardware signal. It could be used to time pulses sent over RF to identify specific symbols. It could also be used to time the response of a specific action like an ultrasonic pulse used to measure distance. The following example shows very accurately measuring an incoming pulse.

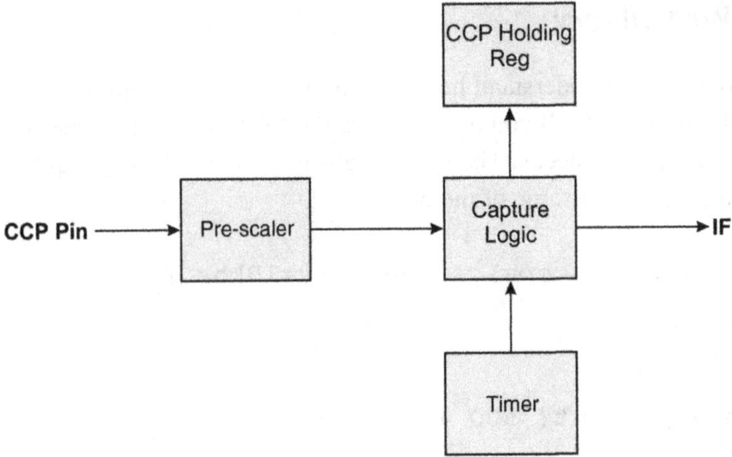

Figure 19.4: PIC MCU CCP module in a capture configuration.

```
enum state = (IDLE, WAIT_FOR_L2H, WAIT_FOR_H2L);
int16 time;

#int_ccp1
void  capure_isr(void) {
   static int16 rise;

   switch( state ) {
      WAIT_FOR_L2H :
         rise= get_capture(1);
         setup_ccp1( CCP_CAPTURE_FE );
         state= WAIT_FOR_H2L;
         break;
      WAIT_FOR_H2L :
         time= get_capture(1)-rise;
         state= IDLE;
         break;
   }
}

void main(void) {
   setup_timer_1( T1_INTERNAL );
   setup_ccp1( CCP_CAPTURE_RE );
   state= WAIT_FOR_L2H;
   clear_interrupt(INT_CCP1);
   enable_interrupts(INT_CCP1);
   enable_interrupts(GLOBAL);
   cout << "Waiting..." <<endl;
   while(state!=IDLE) ;
   cout << "Pulse width is " << precision(1) << (float)time/10 << "us" << endl;
}
```

The **get_capture(1)** gets the value in the holding register for CCP1. The **CCP_CAPTURE_RE** captures on the rising edge. Given that we only looking for two specific interrupts we are careful to clear the interrupts before enabling them. The precision(1) in the cout stream indicates we want to show one decimal place in the floating-point number. The code assumes Fosc = 40 MHz.

In this example we change the edge we are looking for in the ISR to capture the falling edge.

The PIC24 function is very similar. On these parts, instead of the capture peripheral as part of the function call name, it is the first argument. A typical call looks like this:

```
setup_capture( 1, CAPTURE_RE );
```

These parts also have a **CAPTURE_EE** for every edge. By default the timer used is timer 3 but that can be changed to timer 2 like this:

```
setup_capture( 1, CAPTURE_RE | CAPTURE_TIMER2 );
```

In addition the part buffers up the last four captures so you don't need to read them right away.

Some PIC16/18 parts also have some limited ability to select the timer used for a CCP peripheral.

Compare

The compare is kind of the opposite of capture. In this case the CCP register is used to compare against the timer value and when there is a match an interrupt is generated and one of the following functions can be programmed to be performed:

- CCP pin set low
- CCP pin set high
- Toggle CCP pin (only some chips)
- Reset timer to 0 (only some CCP units)
- Start an ADC conversion Chapter 20 (only some CCP units)
- Nothing more than the interrupt.

The first item is a lot like the capability used for the PWM. That is why these functions are related. The fourth item is a way to give a timer a period register for timers that do not have one.

One use for just the interrupt is to force an interrupt a certain time after some event. For example, with a 100-ns timer increment you can program an interrupt to happen 5 ms from now with this code:

```
set_compare(1,get_timer1()+50000;);
setup_ccp1( CCP_COMPARE_INT );
enable_interrupts(INT_CCP1);
enable_interrupts(GLOBAL);
```

Here is code to generate a single very accurate 5-ms pulse high:

```
CCP_1=get_timer1()+10;
setup_ccp1( CCP_COMPARE_SET_ON_MATCH );
while(!input_state(PIN_C2)) ;
setup_ccp1( CCP_COMPARE_CLR_ON_MATCH );
set_compare(1,CCP_1+50000);
```

First we set a nearby arbitrary time to set the pin high, and then wait for it to go high. Then we program the pin to go low 5ms after the time we told it to go high. This all assumes the pin started out low. We assume C2 is the CCP1 pin on this part.

The PIC24 call is similar as would be expected. For example:

```
setup_compare(1, COMPARE_SET_ON_MATCH );
```

In addition some parts allow two times to be set so you can trigger an event on the first time and then the second time is used for the next trigger.

Compatibility Notes

There is nothing in the C standard that deals with any of the capabilities in this chapter.

There is also no standard between C compilers for the PIC® on how to interface to these hardware peripherals. Many compilers do not deal with it at all, leaving the programmer to directly read and write to the SRFs.

These peripherals are very similar between the various PIC® chips so C code that works for one chip will likely work on another. The pins used vary a bit and there are the already identified differences with the 24-bit opcode parts. The number of CCP units also varies between chips. In general this code is easy to port from one chip to another at the C level.

Summary

- Almost all PIC®s have hardware to generate a PWM signal where the frequency and duty can be set.
- A PWM peripheral can be used to generate a clock, or an analog voltage, or to control a motor.

- The capture peripheral can be used to find the exact time according to a timer that a pin changes state.
- The compare peripheral can be used to trigger an event when the timer reaches a certain time. The events can be the change of a hardware pin, resetting the timer, or triggering an ADC reading.
- Interrupts can be generated on all capture/compare events.
- On PIC16/18 class parts the PWM, capture, and compare are combined into a CCP peripheral.
- On PIC24 class parts the PWM and compare functions are combined and the capture peripheral is separate.
- Some parts have a separate PWM peripheral in addition to the CCP/compare to provide an increased-functionality PWM for motor control.

Exercise 19-1

Objective: Learn how to use the PIC MCU PWM and capture units from a C program.
Requires: E3 module, USB cable, PC.

<table>
<tr><th>Steps/Technical Procedure</th><th>Notes</th></tr>
<tr><td>
1. Write a program that uses the E3 buzzer to play Happy Birthday. You can use the notes for the song as:

C D C F E E C D C G F F C C1 A F E D D Bb A F G F

Each note plays a half second and the following are frequencies for these notes:

A 440 Hz
Bb 466 Hz
C 262 Hz
C1 523 Hz
D 294 Hz
E 659 Hz
F 349 Hz
G 392 Hz

HINTS:
• On the E3 board the CCP1 pin is C5. A jumper needs to be installed to connect C5 to the buzzer pin.

Write a program that uses the capture unit to display the frequency of an incoming signal. Measure a square wave 10 times and take the average period to display a frequency on the screen each second.

Use Pin C5, the CCP1 pin, for the input.

HINTS:
• Make sure the signal source on C5 stays above ground (0–5V).
</td><td></td></tr>
</table>

Quiz

(1) A pulse that on the scope looks like a low time of 5 ms and a high time of 15 ms is a signal of what frequency and duty?
  (a) Freq = 50 kHz duty = 25%
  (b) Freq = 50 Hz duty = 25%
  (c) Freq = 50 kHz duty = 75%
  (d) Freq = 50 Hz duty = 75%
  (e) Freq = 5 kHz duty = 15%

(2) For a PWM feeding into a voltage converter, how is the voltage adjusted?
  (a) Change the frequency
  (b) Change the duty
  (c) Change the period
  (d) Change the pre-scaler
  (e) The voltage cannot be changed in firmware

(3) A PWM signal that is a solid low and never pulses is under what condition?
  (a) Broken
  (b) Overflow
  (c) 0% duty
  (d) 100% duty
  (e) If the pin does not change its not a PWM

(4) For the following code, what is the resolution of the duty?

```
setup_timer_2(T2_DIV_BY_1, 99, 1);    //Fosc=40mhz
set_pwm1_duty((int16)50);
```

  (a) 0.01%
  (b) 0.1%
  (c) 1%
  (d) 2%
  (e) 2.5%

(5) For multi-phase motor control in addition to a standard PWM unit you need what?
  (a) Additional hardware to split the signal
  (b) An ECCP unit
  (c) A separate motor-control peripheral in the processor
  (d) Any of the above
  (e) None of the above

(6) Using the capture peripheral to time a single pulse using interrupts, how narrow a pulse can be safely measured with Fosc = 1 MHz?
  (a) 1 us
  (b) 4 us
  (c) 20 us
  (d) 200 us
  (e) 1 ms

(7) An inductive sensor relies on the firmware to pulse a transmitter and then time how long it takes for the signal to be detected by a detector. The target can move over a 50-inch distance. The range in times for the response is up to 100 us. Using the capture peripheral at Fosc = 40 MHz to time the response, what will the accuracy of the measurement?
  (a) +/−0.01″
  (b) +/−0.05″
  (c) +/−0.125″
  (d) +/−0.5″
  (e) +/−1″

(8) Under what circumstances would the capture pre-scaler be used?
  (a) To measure longer pulses
  (b) To measure multiple pulses
  (c) To improve accuracy
  (d) To improve range
  (e) To improve resolution

(9) What can the compare peripheral never do when the timer reaches the target time?
  (a) Set a pin high
  (b) Set a pin low
  (c) Copy the timer value to a holding register
  (d) Reset the timer
  (e) All of the above can be done

(10) What does the following code do for Fosc = 40 MHz?

```
CCP_1=10000;
setup_timer_1(T1_internal);
set_timer1(0);
setup_ccp1( CCP_COMPARE_INT );
```

  (a) Interrupts once in 1 ms
  (b) Interrupts every 1 ms
  (c) Interrupts in 1 ms and then every 6.5536 ms afterward
  (d) Sets the CCP1 high in 1 ms
  (e) Sets the CCP1 low in 1 ms

CHAPTER 20

Analog Techniques

Analog signals have only been briefly mentioned up until now. The signals we have been dealing with are signals (in and out) that are either 0V or Vdd (such as 5V or 3.3V); from a firmware point of view 0 or 1, FALSE or TRUE, and sometimes low or high. When we use the word analog we are referring to signals (input or output) that are a specific voltage between 0V and Vdd, for example 0.5V. In a digital world this will be realized as a 0 but in an analog sense it is 0.5V. This chapter will cover the techniques for handling these analog voltages.

Digital to Analog Conversion

In Chapter 19 we saw a technique for converting a pulse width modulator (PWM) signal to a voltage. This is a crude *digital to analog converter* (DAC). It is crude because it is not calibrated. That is to say it is not easy to generate 3.55V exactly. One method to fix that would be to add a feedback from the analog voltage generated back to the PIC® device on an analog input pin. The PIC® could then read the voltage and adjust the PWM until the voltage is right.

Another, less crude, DAC is referred to as a resistor ladder. When connected to a PIC® it might look like Figure 20.1.

Each digital output when set to 5V will cause a different voltage to be summed into the total and the voltage in the end depends on what combination of pins is set high. A good DAC will use a method like this with very accurate resistors.

Some PIC® devices (but not a lot) have a DAC built in. For those parts it is easy to output an analog voltage. The following code outputs 2.5V on the DAC pin:

```
setup_dac( DAC_VDD | DAC_OUTPUT );
dac_write( 128 );
```

In firmware we deal with voltages as digitized numbers. In this case the built-in DACs are 8 bits and the range on this part is 0V to, in our case, Vdd (due to the **DAC_VDD**). This means 0V is digitized to 0 and 5V (Vdd) is digitized to 255. Each increment is 0.0196V (5V/255). This is our resolution (19.6mV). A 128 then resolves to 128 $*$ 0.0196 or 2.5098V. It should be noted that unless special measures are taken a typical electronics circuit has 5–10mV of noise.

Embedded C Programming. http://dx.doi.org/10.1016/B978-0-12-801314-4.00020-X

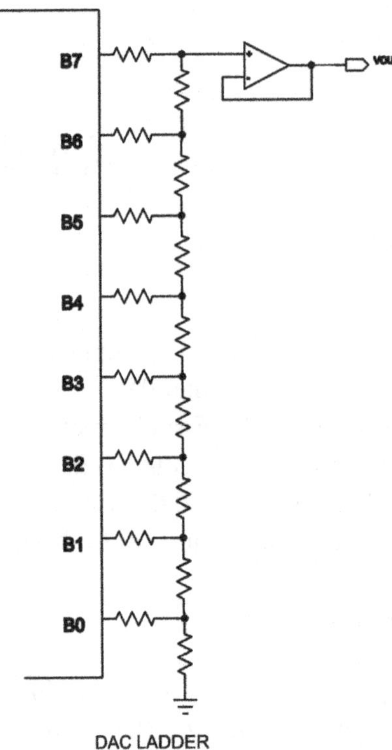

Figure 20.1: Resistor ladder in a DAC configuration.

The DAC also has the capability to use a different upper limit, other than Vdd. A Vref pin on the chip is used as an input that has the analog voltage to use as the top of the range. The new setup call looks like this:

```
setup_dac( DAC_VREF | DAC_OUTPUT );
```

If, for example, there is 2 V on the Vref pin then our new resolution is 0.00078 V and the range is 0–2 V.

For processors with no internal DAC or applications that need better than 8-bit resolution an external DAC can be used. These external DACs are readily available and communicate over a serial bus to the PIC® (Chapter 21 is serial buses). They are commonly used in CD players and MP3 players. A CD does not compress the data and simply has a list of numbers representing analog voltages. The player reads a number (16 bit) and outputs it to a DAC at a fixed rate (around 23 us) to recreate the sound. MP3 players use a compressed format. Sound is combinations of sine waves and given one point on the wave there are only a finite number of positions the next sample could be. This means there do not need to be as many bits to hold the next number; a simple shortened difference number is adequate. Add to this the fact that there are gaps with no sound that can be replaced with codes to indicate how long the silence is, and you can see how easy it is to compress audio.

Some of the 24-bit PIC® chips have a peripheral called a *digital communications interface* (DCI). This is more commonly referred to as a CODEC (coder/decoder). This peripheral uses a standard digital audio format to convert to/from a serial data stream and the digital data representing audio. It is commonly used for digital telephones and other consumer audio that needs digitized audio. There are special DACs that accept DCI format data as input and will give audio out.

Analog to Digital Conversion

Most PIC® processors have an *analog to digital converter* (ADC). In addition they provide multiple pins that can be connected to the ADC through a multiplexer. Firmware must first program the multiplexer for a specific pin and then read the voltage with the ADC. The ADC labels the pins AN0, AN1... and up. The pin-out in the data sheet will show, for example, that pin B3 is also AN5. The compiler simply uses the number part of the AN number (like 0, 1...) to select the analog channel.

The older PIC® processors only had an 8-bit ADC. The standard now seems to be 10-bit ADCs. Some chips have a 12-bit ADC. Even on chips with a 10- or 12-bit ADC you can still use just the top 8 bits to effectively have an 8-bit ADC. Some may do this just to make the code more compatible between chips. Regardless of the number of bits, you can also just shift the ADC value up in a 16-bit word so that the range is roughly the same no matter what the chip. For example, on an 8-bit part the range is 0-FF00, a 10-bit part 0-FFC0, and a 12-bit part 0-FFF0. The decision as to how the result is viewed is made with a preprocessor directive. Here are some examples:

```
#device  ADC=8      // Range is 0-255  on any part
#device ADC=10      // Range is 0-1023  Only for 10 bit ADC parts
#device ADC=12      // Range is 0-4095  Only for 12 bit ADC parts
#device ADC=16      // Range is 0-FF00, FFC0 or FFF0 depending on part
```

The default is ADC = 8 if you do not use this directive at all.

Next you must initialize the ADC peripheral. A typical call to do this looks like this:

```
setup_adc( ADC_CLOCK_INTERNAL );
```

The way the ADC works is that it performs a multi-step (one step for each bit) process to digitize the analog voltage. The above call tells the ADC to use an internal clock, separate from the Fosc, to time each step. The internal clock is around 4 us. Each step takes what is referred to as Tad time, and the full conversion is (bits + 1) * Tad seconds. For a 10-bit conversion with the internal click the time is 44 us. The chip can do a conversion faster and the way to do it is to use the Fosc as a clock. You need to check the data sheet for the minimum Tad (for example 1.5 us) and then based on the Fosc find an available divisor for Tad. On the E3 board chip you can use divisors of 2, 4, 8, 16, 32, or 64. In our case Fosc/64 is 1.3 us and this chip has a minimum Tad of 0.7, so the following could be used:

```
setup_adc( ADC_CLOCK_DIV_64 );
```

This gives us a conversion time of 14.6 us. If time is not critical, using the **ADC_CLOCK_INTERNAL** is safest because it always works on any chip and any clock.

Now we need to set up the analog pins. By default the compiler sets all pins to digital mode so **output_high()** and **input()**-like calls work. Once a pin is set to analog, the digital functions will not work on the pin. Each chip is a little different in the way the analog pins are set up. Some chips allow you to set any combination of pins and some chips only have certain combinations that are allowed. Check the device header file to see what is allowed for your chip. The following sets pins C0 and C1 on the E3 board to analog inputs:

```
setup_adc_ports( sAN4 | sAN5 );      //AN4 is C0 and AN5 is C1
```

Like the DAC, the ADC allows for custom voltage ranges. The ADC defaults to Vss to Vdd (0V to 5V on 5-V parts). The above is the same as:

```
setup_adc_ports( sAN4 | sAN5, VSS_VDD );
```

To use the Vref+ input pin for the top of the ADC scale, use this call:

```
setup_adc_ports( sAN4 | sAN5, VSS_VREF );
```

To specify the bottom as Vref– and the top as Vref+, do this:

```
setup_adc_ports( sAN4 | sAN5, VREF_VREF );
```

The Vref– and Vref+ pins are identified on the pin-out in the data sheet. They are usually another ANx pin so by using them for a reference you lose that analog input pin.

Before we can read the voltage we will need to route the ANx pin to the ADC. This is done like this:

```
set_adc_channel( 4 );
```

After this begins, the internal multiplexer routes the pin AN4 to the ADC. The ADC has a small capacitor in it that will now charge up. When an ADC conversion is done the multiplexer is briefly disconnected and the time it takes for the capacitor discharge is used to digitize the voltage. The reason you need to know all this is that, depending on the input impedance of the analog signal, some time is required for the capacitor to charge after you select the channel. Because the time is dependent on impedance, there is not a fixed time that can be specified. This is only a concern if you switch channels. If you only select one channel and that is the only channel you read then no delay is needed. The full formula for calculating the delay is in the PIC® data sheet. You will need to know input impedance and the operating temperature range. For those without an electrical engineering background, if you have a 1K resistor connected to the PIC® pin and the other end to +5V then that is a 1K input impedance. A direct connect to a power source is near 0 impedance. In general a 10-us delay is good for an input impedance < 10K. Microchip does not recommend using an impedance higher than 10K. In that case they recommend a hardware buffer of some kind be used. Some chips have lower impedance requirements, so you should check the data sheet.

Now all we need to do is the read. The following is a full program to read AN4 and display the voltage on the screen every second.

```
#include <E3.h>
void main(void) {
   int8 adc;
   setup_adc( ADC_CLOCK_INTERNAL );
   setup_adc_ports( sAN4 );
   set_adc_channel( 4 );
   while(TRUE) {
      delay_ms(1000);
      adc = read_adc();
      cout << "Voltage is: " << precision(2) << (float)adc*(5/255) <<
         endl;
   }
}
```

As discussed above, conversion takes a little time. The **read_adc()** will start the conversion, wait for it to complete, and return the result. The **precision(2)** is a **cout** feature that sets the number of floating-point decimal places to two. If you have something else to do you may want to start the conversion and then come back later for the result. Here is how that looks:

```
read_adc(ADC_START_ONLY);
cout << "Voltage is: " ;
adc = read_adc(ADC_READ_ONLY);
cout << precision(2) << (float)adc*(5/255) << endl;
```

More Than 5 V

Improved accuracy for smaller voltages can be achieved with a smaller Vref. What about voltages over 5 V? You cannot use a Vref over Vdd (5 V and on some chips 3.3 V). To read higher voltages the voltage must be scaled down.

One method is to use an op-amp. This was shown in the resistor ladder schematic in Figure 20.1 in a 1:1 voltage configuration. Op-amps can be configured to scale the input voltage down as well.

More commonly a resistor voltage divider is used. The schematic looks like Figure 20.2.

The formula involved is:

Voltage to PIC® = Vin * (R2/(R1 + R2))

For example, if R1 and R2 are both 1K then the voltage to the PIC® will be half the voltage in.

Be aware resistors have a specific accuracy. A 1K resistor sold at 10% tolerance might only be 920 ohms, for example. Extreme temperature changes may also affect the true resistance. More expensive 1% or even 0.1% resistors are available. This needs to be accounted for when determining the ADC accuracy.

It is also a good practice to protect PIC® inputs that come from off-board sources to have some kind of transient suppression part installed to prevent voltage spikes from reaching the PIC®.

Filtering

We already covered the issues in filtering digital signals for noise. There we were only dealing with 1s and 0s. Now our problem is much bigger. Analog filtering can be much more

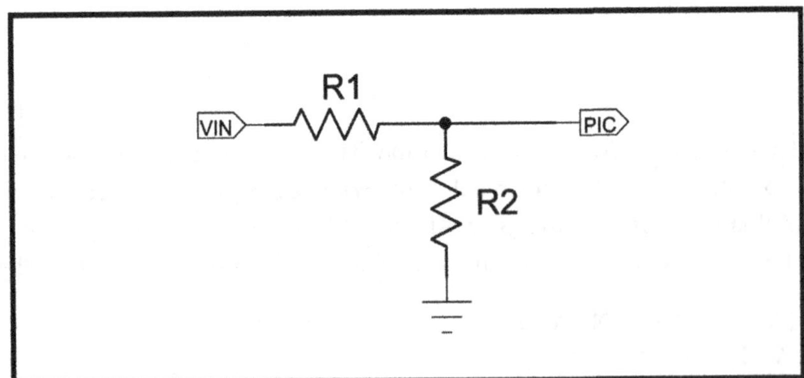

Figure 20.2: Resistor divider schematic.

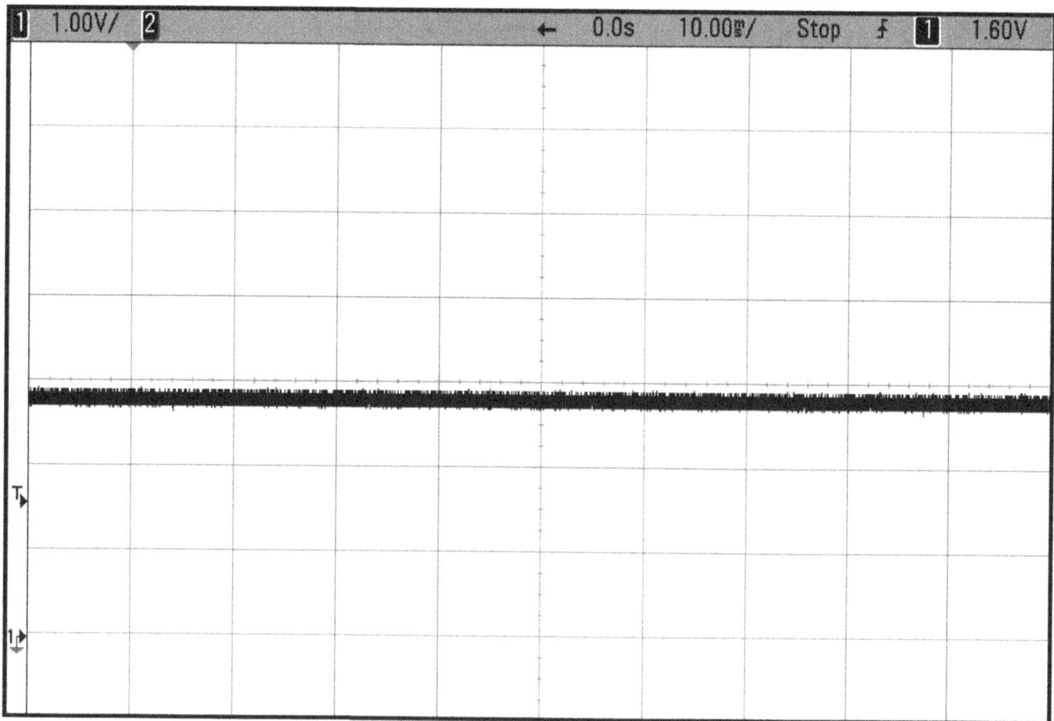

Figure 20.3: Scope trace of a steady analog signal.

complex. Figure 20.3 is a typical analog voltage from a sensor where the voltage from the sensor is not changing.

When we zoom in on this signal to get a better idea of what the sensitive (5 mV at 10 bit) and fast (20 us conversion) PIC® will see, we get Figure 20.4.

Each vertical grid is 100 mV so you can see the signal has 130-mV spikes, a 1-MHz noise of almost 50 mV, and 20 mV of high frequency fuzz. It is clear that, depending on exactly when we take the reading, we might see a variation of at least 20 mV and maybe as high as 130 mV. The first thing a good software person should do is tell the electrical engineer to fix it. Frequently a practical fix is to add a capacitor at the ANx pin of the PIC®. Figure 20.5 shows how two sizes of capacitors affect our noise.

The 47 pf aids on the high frequency noise. It is good at getting out a pickup from the local radio station and other RF noise. The 0.1 uf does a very nice job of bringing the noise down to +/−10 mV. In addition to eliminating the unwanted noise it will distort the actual signal. Figure 20.6 shows the effect of a 0.1-uf cap on a 10-kHz signal. You can see it takes 25 us for a change in the signal to been seen accurately. Depending on the application this may not matter.

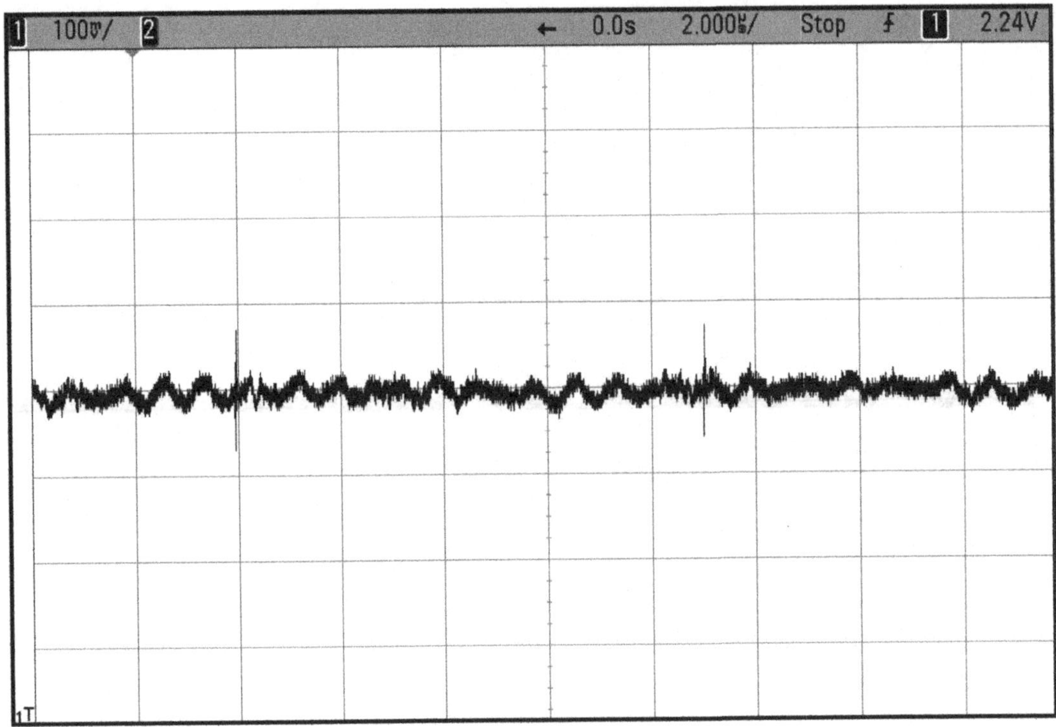

Figure 20.4: Zoom-in of a steady analog signal.

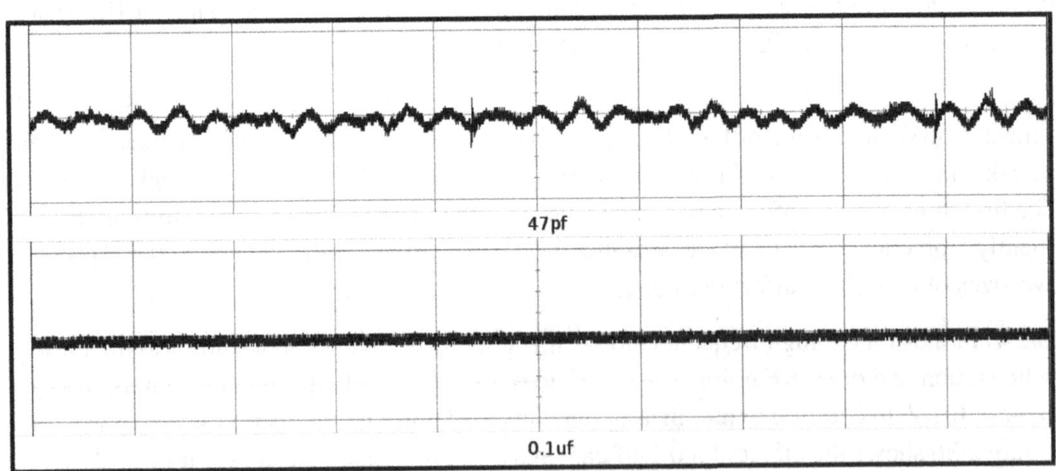

Figure 20.5: Effects of a capacitor on a steady analog signal.

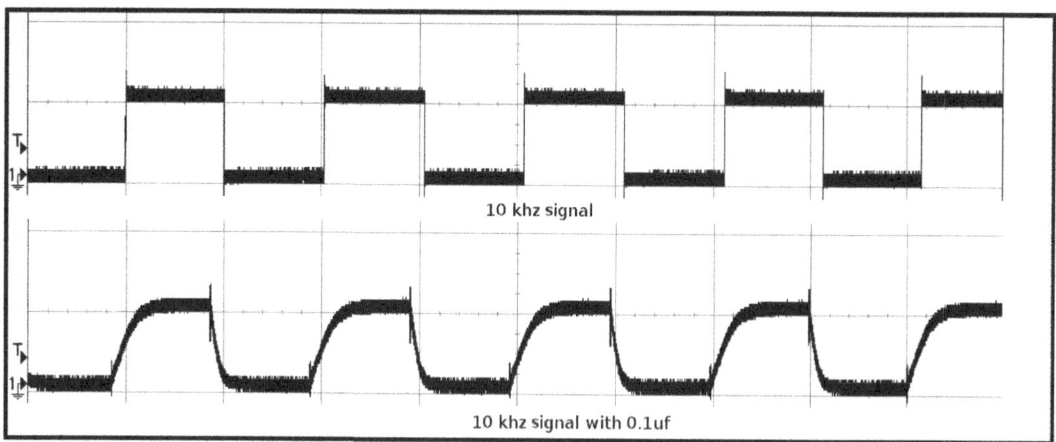

Figure 20.6: Effects of a capacitor on a square wave.

Do not trust these diagrams for your own designs since the pictures will look different depending on impedance.

Once all that can be done is done in hardware, we need to employ filtering in the firmware. For the type of noise we see in Figure 20.4, taking five samples and doing an average will probably get us a reasonable result. The code would look like this:

```
sum=0;
for(i=1;i<=5;i++)
   sum+=read_adc();
result=sum/5;
```

Five samples at 20 us each brings us to 120 us to get an ADC reading. Sometimes this is too long, especially if you are reading multiple channels and also need to delay for channel changes. A common solution is to use a sliding-window scheme to do the averaging. The sampling code would look like this:

```
data[next_ptr]=read_adc();
if(++next_ptr==5)
   next_ptr=0;
sum=0;
for(i=0;i<5;i++)
   sum+=data[i];
result=sum/5;
```

The result would always be the average of the last five samples. You may need to either read five quick samples to fill up the array or add logic to figure out when you start how many samples to average.

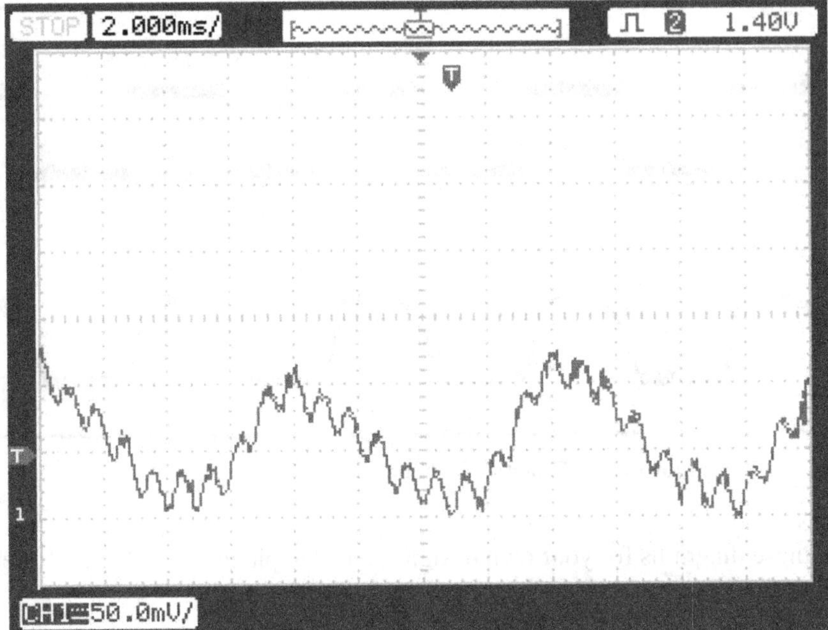

Figure 20.7: Scope trace of a sine wave with noise.

Waveform Analysis

Frequently you will need to do more than read just a voltage. Consider the input signal is a sine wave. You may need to measure frequency, amplitude (peak to peak or RMS), center point, or even distortion. A perfect sine wave is well understood. Figure 20.7 is something more like what may be seen.

It is clear that to measure frequency you cannot just look for a drop in voltage and measure the distance between the first drops. To establish thresholds you need to first determine the peak to peak voltages and then measure frequency with a wide hysteresis. In some cases it might be the higher frequency data you want and the noise is a strong low frequency interference. Be aware sometimes you may also need to measure the noise.

Aliasing

Consider an input signal of 60 Hz and a program that is sampling the signal 10 times a second. Figure 20.8 shows the signal; the dots represent the samples and the lines connecting the dots are how the firmware thinks the wave appears.

What we see is the firmware thinks the waveform is around 12 Hz, not the 60 Hz it really is. This is because of *aliasing*. It is clear from this diagram that you need to make sure you are taking enough samples to capture the fastest waveform you expect to encounter.

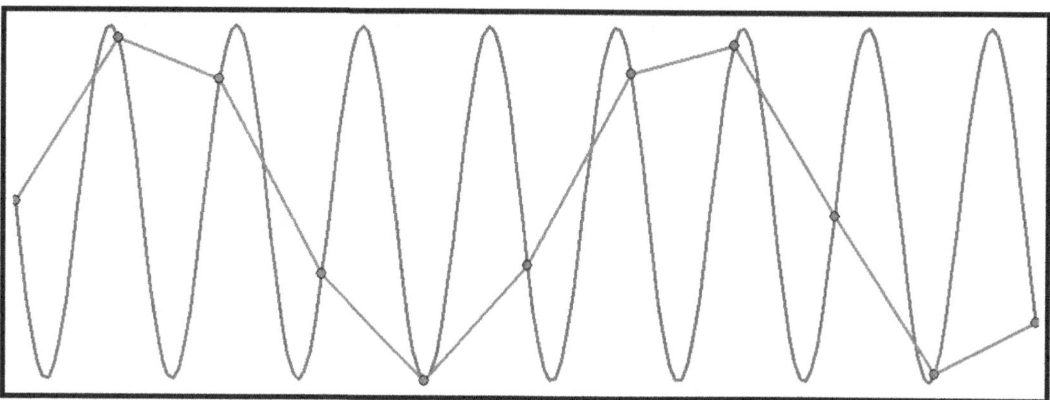

Figure 20.8: Illustration of aliasing due to slow sampling.

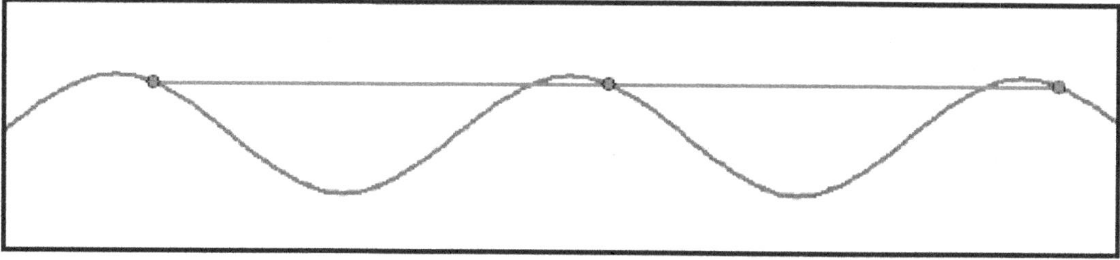

Figure 20.9: Illustration of extreme aliasing when the sample rate is the same as the signal frequency.

It should be pointed out that even with a lot of samples, figuring out the frequency of an analog signal can be a challenge. This is even more true if you have noise in the signal, do not know the voltage range, or if you need to extract frequencies in the presence of other frequencies. These techniques are a more advanced topic which is not covered in this book.

You must also consider aliasing when you have regular noise riding on the signal you need to read. Consider this waveform, where you have a steady voltage out of a sensor but it is picking up a 1-kHz noise by the time it reaches the PIC®. The firmware is doing 1000 samples per second to read the signal. See Figure 20.9.

What has happened is the firmware thinks the signal voltage is higher than it really is because it happens to sample the noise only at its high points. You can see the same problem happens with 500 samples per second and you can get other distortions at other frequencies.

As you are thinking, "what are the chances the sample rate and noise are the same frequency?" consider the possibility the noise is being caused by the firmware. For example, a program that operates on a 10-ms loop where 100 times a second the firmware does eight things. For instance, to send data out, engage a relay, or change the speed of a motor, all of which can cause a noise pulse. Another thing the program does is read the ADC. The ADC reading may always (or frequently) be done in the presence of an unusual bump in the noise floor.

Working in Your Sleep

It should be clear by now that with analog inputs, noise is a big issue. One of the big sources of noise around a PIC® device is the oscillator that supplies the clock to the processor. The PIC® sleep command will shut off the oscillator, significantly reducing the noise level. In addition to the watchdog timer (WDT), interrupts can wake the chip out of sleep and the ADC has a completion interrupt that can be used for just this purpose. Be aware, to use this feature you must run the ADC clock off the internal ADC clock because the chip clock is turned off. Here is what the code looks like:

```
disable_interrupts(GLOBAL);   // If it was enabled
enable_interrupts(INT_ADC);
adc_read(ADC_START_ONLY);
sleep();
value = read_adc(ADC_READ_ONLY);
disable_interrupts(INT_ADC);
enable_interrupts(GLOBAL);   // If it was enabled
```

You want the global interrupts turned off so the interrupt service routine (ISR) function is not actually called (in this case we do not have one). The setting of the interrupt flag (IF) is enough for the chip to wake up. Be aware that the timers are all stopped while the chip is asleep. Also take note that some chips have special sleep modes that control which peripherals are shut down during sleep. If available, those options are passed as an argument to **sleep()**. With no arguments, the function shuts down as much as possible.

Voltage Reference

As covered above, an external Vref can be used to identify the maximum voltage. One reason to use this even if your maximum voltage is Vdd is that Vref may provide a stable known voltage. There are at least two problems with Vdd. First, some battery-operated devices might actually have a sloppy Vdd where the voltage might be anywhere from 3 V to 5 V depending on the battery health. The second problem is that with common voltage regulators you can see a 5% variance of the voltage depending on the part and the temperature.

As an example, consider a humidity sensor that outputs 1 V for 0% humidity and 4 V for 100%. If you have a Vdd of 4.5 V but in firmware assume it is 5 V, then for a 50% (2.5 V) humidity reading you get an ADC value of 141 and you assume that is 2.77 V, or 59% humidity (9% error).

On the other hand you may have a temperature sensor that outputs 0.01 * Vdd for every degree. As long as that sensor's Vdd is the same as the PIC® Vdd then the actual Vdd value doesn't matter.

One more example. A battery-operated device uses the ADC to monitor battery voltage. When the voltage reaches 3V the device lights the LO BAT LED. If the battery voltage directly powers the PIC® Vdd then the PIC® will never think the battery is low because the AN pin will always equal Vdd. That will always read 255 from the ADC.

To get a sense of real voltage (not based on Vdd) we need a voltage reference. There are voltage reference parts you can buy for this purpose. Some PIC®s have a peripheral for generating a precision voltage reference for you. Some even allow you to route the voltage to the ADC directly. These peripherals usually have a few voltages you can select from. There is a lot of variance between parts on this peripheral so you do need to check the header file for the options on your specific part. In the case of the E3 board, there are three voltages available: 1.024, 2.048, and 4.096V. The setup looks like this:

```
setup_vref (VREF_4v096);
```

To route this voltage to the ADC use this:

```
setup_adc_ports( sAN4, VSS_FVR );
```

FVR is what this part calls the internal voltage reference. Again there are differences in each part concerning the voltage reference.

Always check the accuracy as a percentage for the voltage reference you are using. There is a wide range of accuracies and you will find the references built into the PIC® are not the most accurate.

Comparator

Many PIC® processors have a comparator peripheral. Unfortunately there is a lot of variation between the chips as to the specifics of how the peripheral works. There is a type of operational amplifier in the EE domain called a comparator and the PIC® peripheral works much like it. The electrical symbol looks like that shown in Figure 20.10.

The way it works is if the + input voltage is above the − input voltage then the output voltage is a high (two analog in, one digital out). The way this is usually used is a reference voltage is put on the − and then your sensor input is the +. You then get a high signal out when the + signal rises above the threshold.

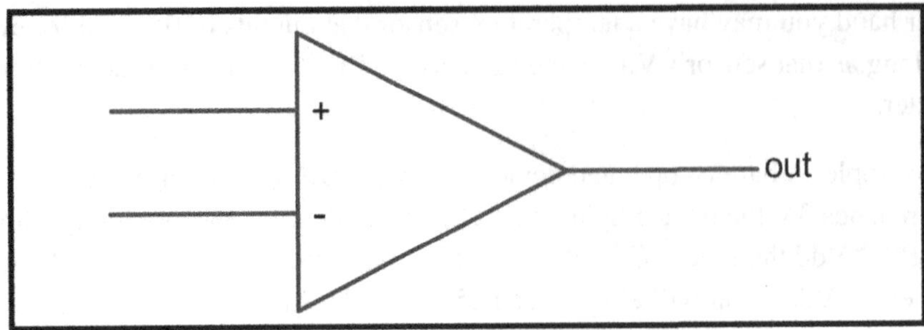

Figure 20.10: Comparator schematic symbol.

For the PIC® this output signal can generate an interrupt and/or be routed to an output pin. For PIC® chips with a Vref generator you can route that voltage to one of the comparator inputs. This can be used as a low voltage alarm or any number of detectors that are based on an analog voltage.

The PIC® device used on the E3 board has two comparators and both look similar to the simplified block diagram shown in Figure 20.11.

Notice there are multiplexers that allow the programmer to select the specific inputs to the comparator. Be aware that for the pins you must use the **`setup_adc_ports()`** function call to set the pins to analog mode (as opposed to the default digital).

The output has a selection where you can get the output as is or you can invert the output. You can get an interrupt if enabled, and optionally enable an output to pin C4. There is also an optional connection to the PWM peripheral. This is used for a fast shutdown of the PWM and can be used in motor control.

Here is code that shows a warning if the voltage on C2 drops below a reference level set in hardware:

```
#int_comp
void comp_isr(void) {
   cout << "Warning: C2 low" << endl;
}

void main(void) {
   setup_adc_ports( sAN6 );
   setup_comparator( CP1_C2_VREF | CP1_INVERT );
   clear_interrupt(INT_COMP);
   enable_interrupts(INT_COMP);
   enable_interrupts(GLOBAL);
}
```

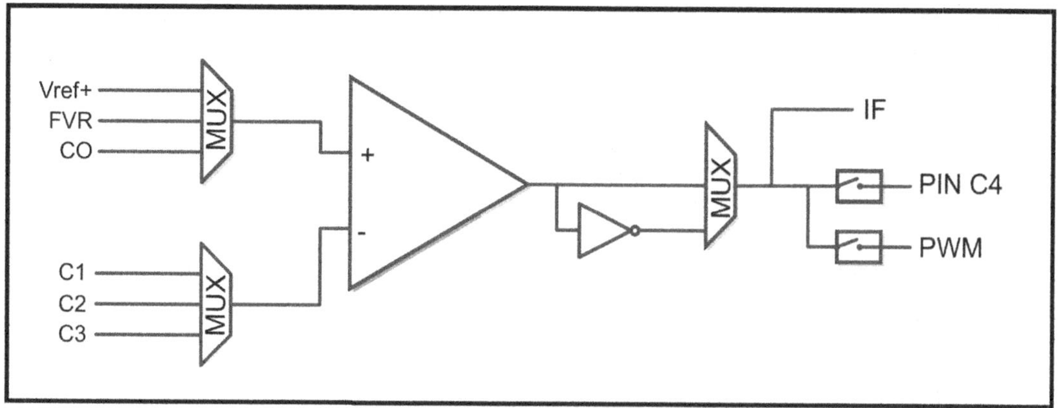

Figure 20.11: Example PIC® MCU comparator functional diagram.
MUX = multiplexer.

The **C2_VREF** indicates C2 is assigned to the − and the Vref pin to the +. To interrupt when C2 drops below VREF we need to invert the output.

C2 could be connected to a sensor that detects power before the regulator/capacitor that supplies Vdd. In this case software could get an early warning that power is about to fail. If the Vdd capacitor is large enough then there could be enough time to save the system state in data EEPROM (electrically erasable programmable read only memory). For example, a control unit that fills a tank with liquid by running a pump for a given amount of time. If power is lost, a flag indicating the tank needs filling along with how much longer the pump needs to be run could be saved in data EEPROM just before power fail. When power is restored the processor can continue where it left off.

Voltage Detect

Some processors have hardware to detect when the voltage drops below a set threshold. Many chips have what is called a brownout detect. This is set up with fuses and in all cases causes the chip to be held in reset while the voltage is too low. This prevents the processor from acting insanely when it is run at a too low a voltage. Here is how this could be set up to stop running when the voltage drops below 4.2V:

```
#fuses  BROWNOUT, BORV42
```

The voltage levels allowed vary with each chip.

Some chips have a special interrupt that can be set to trigger as Vdd drops. This is called low voltage detect, sometimes high/low voltage detect. This may allow for some preparation for a loss of power. Usually this is not enough to save data to EEPROM, however. It may be enough to do a safe shutdown of items being controlled, like motors. The following is example code:

```
setup_low_volt_detect( LVD_38 );     // 3.8V
enable_interrupts( INT_HLVD );
enable_interrupts( GLOBAL );
```

The interrupt is called HLVD (high/low voltage detect) because some chips also allow a trigger for too high a voltage.

Compatibility Notes

There is nothing in the C standard that deals with any of the capabilities in this chapter.

There is also no standard between C compilers for the PIC® on how to interface to these hardware peripherals. Many compilers don't deal with it at all, leaving the programmer to directly read and write to the SRFs.

These **setup_adc_ports()** and **setup_comparator()** arguments vary a lot between PIC® chips so the header file must always be consulted concerning these arguments. The voltage reference and voltage detect also vary a lot between chips and some chips do not have those capabilities at all.

Summary

- Digital to analog converters (DACs) convert from a digital number to a voltage.
- A few PIC® devices have internal DACs and external DACs are quite common.
- Some PIC®s have a DCI unit for dealing with standard digitized audio.
- Most PIC® chips have analog to digital converters (ADCs) to convert a voltage input to a numeric value.
- PIC® ADCs have a multiplexer to select one pin from a number of analog input pins.
- The resolution on PIC® ADCs is usually 10 bits; however, some PIC®s have 8-bit and 12-bit resolution.
- Pins must specifically be set to the analog mode, as opposed to digital.
- A voltage reference may be used for the high and optionally low voltage range used by the ADC to improve resolution over a limited range.
- Resistor dividers may be used to measure voltages higher than the chip Vdd.
- Frequently some kind of filtering must be done on analog signals to discard noise.
- Care must be taken so the sample rate does not distort a signal due to aliasing.
- The sleep function can be used to quiet the chip for sensitive sampling.

- Some PIC® processors have an internal voltage reference generator that can be used for an accurate voltage reference.
- Many PIC® processors have a built-in voltage comparator that can be used to trigger an event when one analog voltage rises above another voltage.
- Many PIC® chips allow for a low voltage detection by either causing a reset or causing an interrupt.

Exercise 20-1

Objective: Learn how to use the PIC® MCU ADC from C.
Requires: E3 module, USB cable, PC, 10-uf capacitor, 1-uf capacitor, 100K 1% resistor, 9V battery, and clips.

Steps/Technical Procedure	Notes
1. Write a program that displays a chart on the PC screen showing the position of the pot on the development board. As the pot is turned, the voltage on the chart should show the change. The E3 board pot is connected to PIC® pin AN0 (aka A0). The E3 pot ranges from 0 to 5V. **HINTS:** • The charting capability of the development tools was introduced in Exercise 11-1.	
2. Write a program that shows on a chart the discharge curve of a 10-uf capacitor through a 100K resistor. The capacitor should be connected to an AN pin on the + side and the − side should go to ground. The resistor should be connected to the same AN pin and the other side to ground. **HINTS:** • In the digital mode, output a high to the capacitor pin for a couple of seconds to charge the capacitor up. • After charged, switch to analog mode and begin sampling the voltage at the AN pin as the capacitor discharges through the resistor.	

(continued)

Steps/Technical Procedure	Notes
3. Write a program that figures out the capacitance of a capacitor. Display the result in microfarads on the PC screen with one decimal point. Test with a 10-uf and 1-uf capacitor. The time for a capacitor to discharge through a resistor to 67% of the fully charged voltage is given by: seconds = farads * ohms A farad is 1,000,000 microfarads (uf). **HINTS:** • Discharge using the method in exercise part 2, and time how long it takes to get to 67% of Vdd. • Assume the resistor is exactly 100K (because it has 1% accuracy) and using the measured time calculate capacitance. • Do not expect the calculated values to match the value stamped on the captivator. The parts as sold are sometimes as bad as +/−20% of the value.	
4. Write a program that measures to the highest accuracy possible the voltage of a 9-V battery. Display the voltage to three decimal places and also display the measured noise picked up as a +/− voltage number to three decimal places as well. Numbers should be displayed each second. The E3 board has a two-resistor voltage divider on pin 7 of the header going to C7. The resistor to pin 7 is 3K and the resistor to ground is 1K. Both are 10%. The processor on the E3 board can be programmed to output a Vref on pin C2. **HINTS:** • Two programs are required. First a program to figure out the actual divide ratio on the pin 7/C7 input is needed. The Vref generator in the chip can be used to output a known voltage. Jumpered to the voltage divider the exact ratio can be determined. • For highest accuracy the 10-bit mode on the ADC needs to be used. • Use a sliding-window scheme to filter the data and use the same array to calculate noise based on the voltage variation in the collected numbers. • Do not attempt to use the `sleep()` mode on the E3. It will disrupt the PC USB communication.	

Quiz

(1) If an ADC or DAC has a 12-bit resolution from Vss to Vdd and Vdd is 5 V then what is the voltage difference from a numeric 10 to a 12?
   (a) 0.244 mV
   (b) 0.488 mV
   (c) 2.44 mV
   (d) 4.88 mV
   (e) 5.37 mV

(2) What is the formula to find the resolution of an 8-bit ADC or DAC?
   (a) 28
   (b) Vdd/256
   (c) Vdd/255
   (d) Vdd/8
   (e) It is not the same formula for an ADC and DAC

(3) What happens if a digital voltage is applied to an analog pin and an 8-bit **`read_adc()`** is done on the pin?
   (a) It always reads 0
   (b) It will read either near 0 or near 255 depending on the signal state
   (c) It will read 128
   (d) You should get a compile error
   (e) The results are undefined

(4) Of the following, what is the fastest frequency that could be measured using only an analog input on a typical Fosc = 40 MHz processor?
   (a) 40 MHz
   (b) 10 MHz
   (c) 4 MHz
   (d) 400 kHz
   (e) 40 kHz

(5) If a processor has Fosc = 48 MHz and the minimum conversion time is 33 us for the full 10-bit ADC, which of the following options will give us the fastest analog conversion and still work?
   (a) **`ADC_CLOCK_DIV_64`**
   (b) **`ADC_CLOCK_DIV_32`**
   (c) **`ADC_CLOCK_DIV_16`**
   (d) **`ADC_CLOCK_DIV_8`**
   (e) **`ADC_CLOCK_INTERNAL`**

(6) For a 10-bit ADC set with a range of VSS_VREF and the VREF is an external 3.5-V reference, what does a return value of 100 from `read_adc()` mean?
(a) 0.035 V
(b) 0.350 V
(c) 1.961 V
(d) 1.367 V
(e) 1.373 V

(7) A 5-V processor with a 10-bit ADC needs to read a battery voltage that could be as high as 18 V. Using a resistor divider, which of the following combinations will give the best accuracy and still stay in the safe range?
(a) 6K and 2K
(b) 5K and 2K
(c) 4K and 2K
(d) 10K and 4K
(e) K and 4K

(8) An application needs to digitize a voice input and save 5 seconds, worth of the voice. The desired quality dictates a 4K sample rate and an 8-bit ADC. How much memory will be required to save the voice and how fast must the average write time for a byte be?
(a) 4096 bytes and 3.9 ms/byte
(b) 4000 bytes and 244 us/byte
(c) 16,384 bytes and 4.9 us/byte
(d) 20,000 bytes and 4.9 us/byte
(e) 20,000 bytes and 250 us/byte

(9) A cash register program has an interrupt trigger when the Vdd reaches 3.5 V. Of the following, which is a good fit for what could be done in the ISR?
(a) Lock the cash drawer
(b) Show a warning on the screen to alert the operator of a pending power fail
(c) Close the current transaction and print the receipt
(d) Send all the day's transactions to a host computer
(e) Send an e-mail to the power company requesting power restoration

(10) What is a good reason to use a voltage reference?
(a) When the application requires a DAC
(b) To increase range
(c) To improve resolution over a more limited range
(d) When the analog voltage needs to be known in true volts
(e) c and d

CHAPTER 21

Internal Serial Busses

Serial busses send and receive data 1 bit at a time. Serial busses reduce the complexity of a hardware design and free up pins on the processor. All this comes at the expense of time. A device that has a parallel interface can clock in 8 data bits with one quick pulse of a pin. For a serial bus each bit must be sent separately and this takes time.

This chapter will cover the SPI and I^2C protocols since these are most popular protocols for processor communication to a nearby device.

Internal serial busses are defined as communicating over short distances, frequently less than a foot. A variation of RS-232 for internal serial busses is frequently used for processor-to-processor communication and is covered in Chapter 22 because, from a programming point of view, it looks similar to long distance external communication.

Serial Peripheral Interface

The *serial peripheral interface (SPI) protocol* is a simple serial interface that can be used to communicate with devices that do not have a microprocessor. It is very easy with shift register logic and gates to implement SPI on devices such as a data EEPROM (electrically erasable programmable read only memory), external ADC (analog to digital converter), or simple sensors. Typically SPI uses two or three wires plus an additional wire for every device on the same bus. The extra wires are to select the device you want to talk to.

The SPI bus is a very loose standard. Different manufacturers have developed their own version of what they thought was best and programmers are left to figure out what needs to be done for each device. SPI is a synchronous protocol. This means a clock is transmitted along with the data. Figure 21.1 shows an example hardware connection schematic.

SPI has a master and slave. In Figure 21.1 the PIC® device is the master and the temperature sensor is the slave. The master sends out the clock pulses to the slave to indicate data transfer. The example in Figure 21.2 shows what the data transfer over the four wires looks like.

We have an 11-bit transfer. Each data bit is transferred when the clock goes from low to high. First the PIC® sends out a command of 4 bits. 1001 means read sensor #1. Then the slave responds with 7 bits. In this case 1000110 means 70 degrees. The slave select (SS) enables the

Embedded C Programming. http://dx.doi.org/10.1016/B978-0-12-801314-4.00021-1

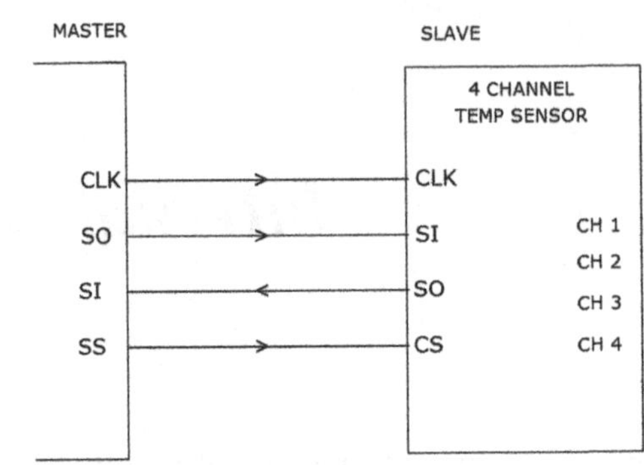

Figure 21.1: SPI bus master/slave connection diagram.

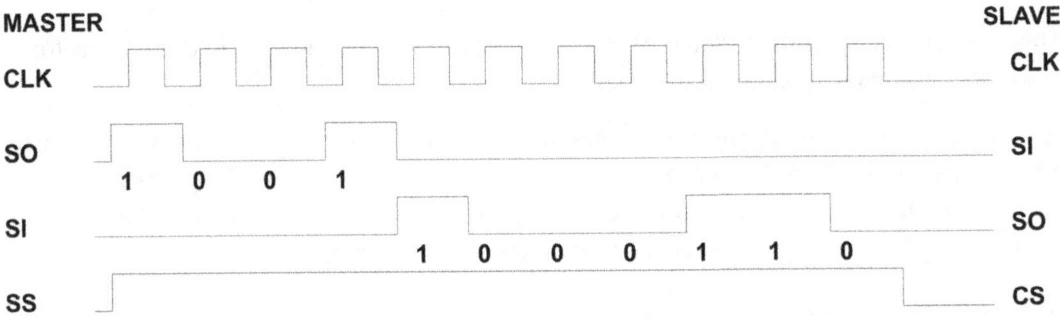

Figure 21.2: SPI bus timing diagram.

slave device during the entire transfer. The slave CS (chip select) ignores the bus while it is low.

The number of bits for a command and response varies depending on the slave device.

Many SPI protocols will use commands that start with a 1 bit. They will then ignore extra clock pulses with data bits as 0. This helps if the data sizes are different in the two directions. Remember, for every clock pulse data can be transferred both directions.

Let us look at some code:

```
#define  TS_CLK PIN_B0
#define  TS_SO  PIN_B1
#define  TS_SI  PIN_B2
#define  TS_SS  PIN_B3

cmd=0b10010000;
data=0;
```

```
output_high(TS_SS);
for(i=1;i<=4;++i) {
   output_bit( TS_SO, shift_left(&cmd,1,0));
   output_high(TS_CLK);
   output_low(TS_CLK);
}
output_low(TS_DI);
for(i=1;i<=7;++i) {
   output_high(TS_CLK);
   shift_left(&data,1,input(TS_SI));
  output_low(TS_CLK);
}
output_low(TS_SS);
```

The command is padded with 0s in the low bits so we could use the **shift_left()** function to easily shift out the bits one at a time. **shift_left()** is again used to shift in the data 1 bit at a time. This code does not show any delays to hold the clock up for a period of time. For many SPI devices the PIC® cannot go fast enough to break the timing rules of the slave. Always check the slave data sheets, however, to make sure the timing is respected. They will indicate the minimum time the clock can be low and high and the minimum time the data can change before and after a clock change.

Although the above code clearly shows the protocol specifics in C, the CCS C compiler has an easier way to do this code. The following uses the SPI library:

```
#use spi(  master, bits=11, clock=PIN_B0, so=PIN_B1, si=PIN_B2,  \
      enable=PIN_B3, sample_rise, enable_active=1 )
...
data = spi_xfer(0x10010000000);
```

The **spi_xfer()** sends out and receives 11 bits concurrently. That is why we padded 7 bits to the command. Some SPI devices transfer data in both directions at the same time. Sometimes this means the master is sending a command and getting a response for the previous command at the same time.

The **#use spi** specifies the data is sampled on the rising edge (the default is falling edge) and it needs to indicate the enable pin (aka SS or CS) is active high. The default is active low because most slaves are selected when the CS goes low.

The clock rate and many more SPI protocol specifics can be specified in the **#use spi**.

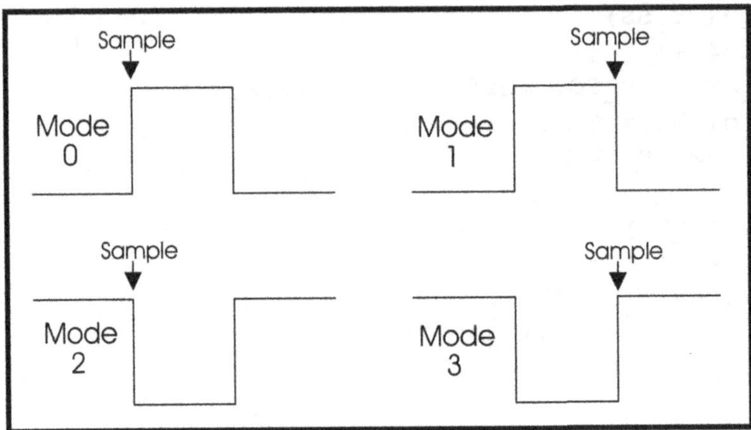

Figure 21.3: SPI modes chart.

SPI Modes

There are four combinations of the edge the data is sampled on and the idle state of the clock. These four combinations are commonly called SPI modes and the modes are numbered 0–3 or sometimes in binary 00–11. The **`#use spi`** directive allows you to simply use **`mode=n`** where **`n`** is the mode in decimal or binary to set up the protocol for your device. The chart in Figure 21.3 shows all four modes.

Hardware SPI

The **`#use spi`** works on any pins you specify. It generates code like that shown above, that bit-bangs the pins to do what needs to be done. The PIC® processors have a built-in peripheral to do SPI communication. If the pins you select are connected to the SPI module (called SSP) then the hardware SPI is used. Another way to get the hardware SPI is to just identify the hardware module (like SPI2). Here is an example of hardware SPI:

```
#use spi(master,SPI1,bits=11,enable=PIN_B3,sample_rise,enable_active=1 )
```

For a master there is no predefined hardware enable pin (the SS pin is only used if the PIC® is a slave). The CLK, SI, and SO pins need to be connected to the predefined SPI1 pins on the chip being used.

Also be aware that when a hardware SPI is used the actual data sent is always a multiple of 8 bits. The outgoing data is padded with zeroes as the first bits out.

Multi-drop SPI

When there are multiple slaves on the same SPI bus this is referred to as a multi-drop situation. Figure 21.4 is a two-device schematic:

You can see there is an extra SS/CS wire for every slave device. Only one device is active at a time. When a device's CS pin is inactive the part does not drive the SO pin, it is high impedance so other devices can use the wire.

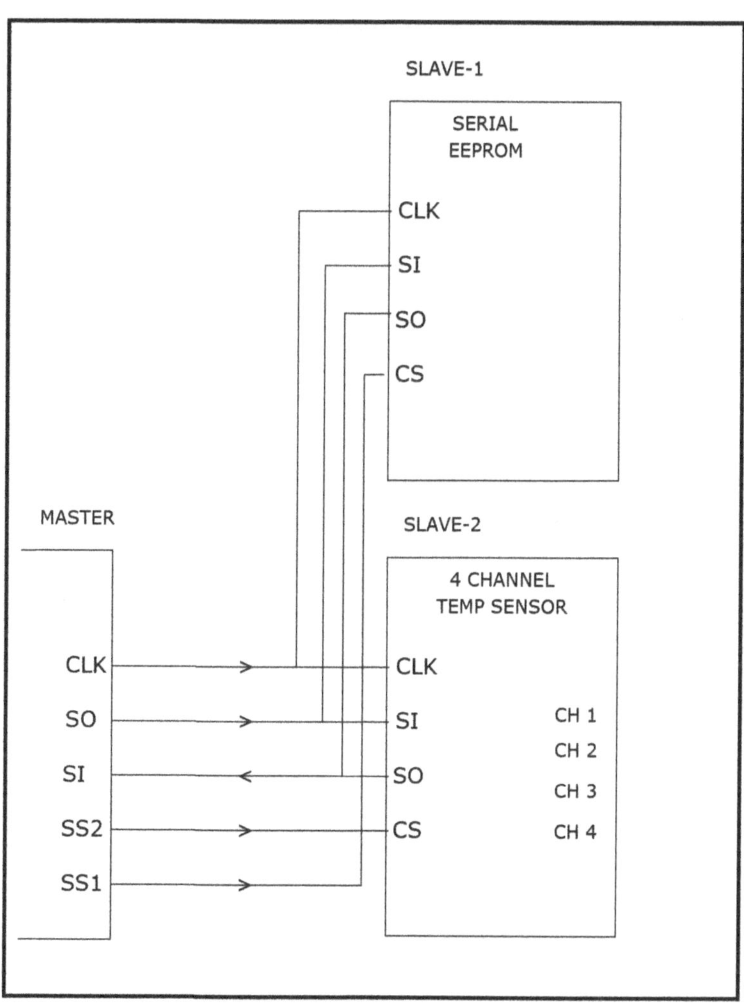

Figure 21.4: SPI connection diagram for two slaves.

Fewer Wires

If there is only one slave device sometimes the SS/CS wire can be skipped and the slave will simply tie CS to Vdd (or Vss for active low CS). This saves a wire. Make sure from the data sheet that the slave supports this, however. Some slaves use the transition from inactive to active on the CS pin to reset the shift register to accept a new command.

Another optimization that may be done is to eliminate one direction of data transfer. For example, a DAC (digital to analog converter) frequently only accepts incoming data and has no reason to send data back to the master. In this case the SI/SO wire can be eliminated.

Finally if the protocol does not have concurrent data transfer then the same wire can be used for both directions. For example, in our first example above a command is sent and then data is returned but not at the same time. For such a case, Figure 21.5 is a schematic showing a simple two-wire SPI.

Noise

Noise is a big problem in SPI. This is in part because the slave devices accept a high frequency clock, and they also accept as a clock a strong noise pulse. If you are transferring, say, an 8-bit word to the slave and in the middle it takes a noise pulse as an extra clock then all the data from that point on is shifted. Not much in firmware can be done to solve this.

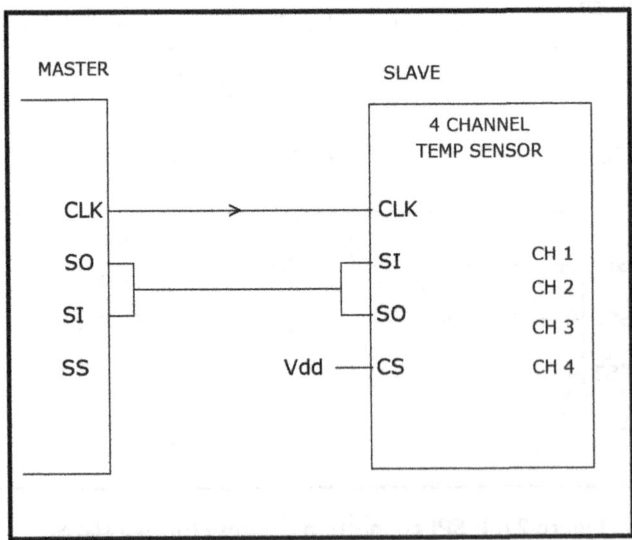

Figure 21.5: SPI 2-wire connection diagram.

If your slave device resets the shift registers on some action, like toggling the CS pin, then doing this on every transfer will reduce the damage. If you have a slave that ignores zero data pulses until it gets a 1 as the start of a command then this trick can be used to ensure synchronization:

```
spi_xfer( 0 );  // sends out 11 zero data bits
data=spi_xfer( 0b1001000000 );  // Sends command 1001 and gets 7 bit reply
```

The first **spi_xfer()** clears the shift register in case it is already starting shifting in a command.

Framing Signal

For very high speed SPI an additional wire is used to help synchronize the data. Only the PIC24 class parts support this extra wire. They are the only parts that can do super high data transfers. These kinds of transfers are used for real-time digitized voice, video, and other high data throughput communication.

The way it works is the new signal (called frame signal) is pulsed at the start of what is called a frame. A frame might be a single byte, a 16-bit word, or up to 127 bytes. This solves the problem where a noise pulse gets all the future communication shifted by one. There is a positive synchronization every frame. Frequently when the framing pulse is used the clock runs free, never stopping because data is always preceded by a frame pulse. The frame pulse is usually sent by the device sending data. That is not always the master.

Being a Slave

Sometimes a PIC® needs to be an SPI slave. This may be done when SPI is used for chip-to-chip communication (although RS-232, below, is a better choice). SPI is also sometimes used for device-to-device communication or to emulate an SPI device for testing.

It is best to use hardware SPI when a slave. This is because you can get an interrupt when data comes in. The alternative is to always be waiting for data and you may not be able to handle consecutive transmissions without a deliberate gap in time.

The complication with hardware SPI is the data transfers are multiples of 8 bits. Going back to our first example, after the first 4 bits the slave responds. With hardware SPI we won't even know the master is sending anything until 8 clocks are out. For this example we will change the protocol to an 8-bit command and 8-bit response. Here is what the interrupt service routine (ISR) looks like:

```
#int_ssp
void spi_slave_isr(void) {
   unsigned int8 cmd;

   cmd = spi_xfer ();
   if(cmd!=0)
      spi_prewrite( get_temp(cmd & 3) );
}
```

On the PIC16/18 parts the SPI interrupt is **int_ssp**; on the PIC24 parts it is **int_spi**. The call to **spi_xfer()** gets the 8-bit command and nothing is sent out during those 8 clock pulses. The **spi_prewrite()** queues up data to be sent out with the next 8 clocks. We will get another interrupt after the next 8 clocks come in and that data goes out. In that case the cmd from the master will be 0 so we do not preload the outgoing queue in that case.

I²C

SPI can be cumbersome from a hardware point of view when you have lots of devices on a bus. Consider a car radio where you have a device that controls tone, another that controls volume, double those for each channel (right and left), and then add an LCD controller for the display and keypad controller for the user interface. Do not forget a real-time clock, frequency synthesizer for the radio, and the interface to the auto bus. By this time you have piles of CS lines all going to the processor. One solution is to have a separate SPI device on another SPI bus that accepts a command to select a device and that device controls all the CS wires. What is an easy way to talk to a couple of devices becomes a mess with lots of devices.

The I²C protocol solves this problem and more. I²C is always exactly two wires for as many devices as you want (or at least around 100). I²C is similar to SPI but with some specific rules that make it cleaner for many devices and more robust for low data rate communication. The key facts about I²C are as follows:

- All data transfers are multiples of a byte (8 bits).
- Each slave on the bus has a unique address.
- Every byte transferred is specifically acknowledged by the receiver. There are some device-specific exceptions to this rule.
- All transfers begin with a special start condition and end with a special stop condition.
- The first byte after a start condition always comes from the master and has a 7-bit address for the desired slave and a 1-bit direction indication to indicate who sends the remaining data bytes (master (0) or slave (1)).
- The data direction can be changed in the middle of a transfer by the master by sending a new start condition followed by a new address/direction byte.

- The slave can slow down the master if the slave is busy using a technique called clock stretching. The slave simply drives the clock line low, preventing the master from raising it.
- There are two accepted speeds, fast at 400K bits per second and slow at 100K bits per second.

The I^2C hardware names the two wires SCL (clock) and SDA (data). Both are open collector outputs from all devices including the master. The devices either drive a pin low or let it float. The bus must have pull-up resistors on each wire and that is what causes the wires to go high. This way no bus user can ever cause damage by driving the bus in the opposite direction of another user. This is why the clock stretching mentioned above works. When the master wants to make the clock high he just lets go of the SCL. If it does not float high, then the master knows a slave is stretching the clock and the master waits. See the schematic in Figure 21.6.

For normal data the data line changes when the clock is low. It is clocked in on the falling edge of the clock. The start and stop conditions change the SDA line while SCL is high. This way they are unique conditions that can be detected. The start and stop conditions look like Figure 21.7.

The first byte after a start condition looks like Figure 21.8.

The slave address may be fixed in the device or it may be a combination of some bits fixed and others set by a combination of fixed bits and pins. Figure 21.9 is a pin-out for a 2401 serial EEPROM.

The address of the part is 1010xxx where xxx is set by pins A0, A1, and A2. These pins are grounded for 0 or pulled to Vdd for 1. If the A0 is connected to Vdd and the other two to

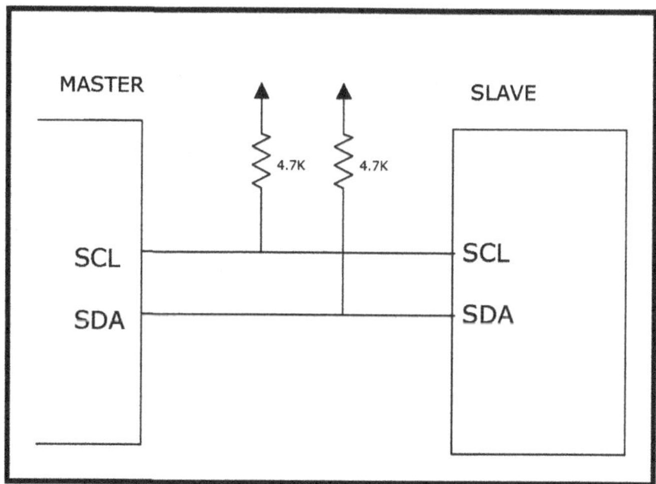

Figure 21.6: I^2C bus master/slave connection diagram.

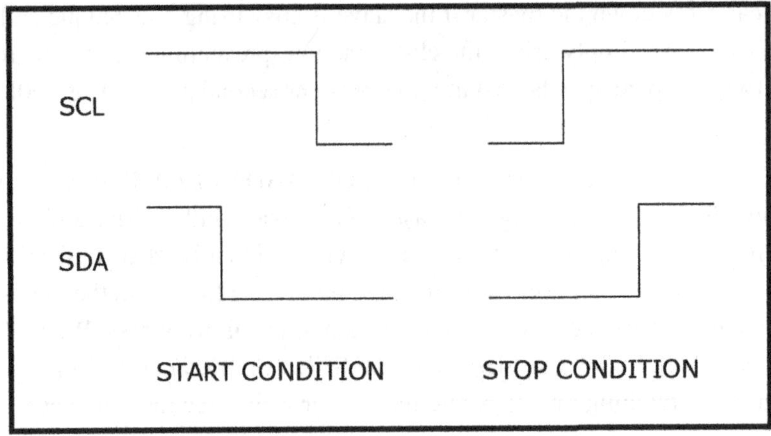

Figure 21.7: I²C bus special condition timing diagram.

```
#use i2c(master, sda=PIN_B1, scl=PIN_B2, fast)
```

Figure 21.8: First I²C byte mapping.

ground then the address is 1010001. This means we can have up to eight of these devices on the same bus.

Now if we want to send data to the slave, the byte after the start condition would be 0xA2 and if we want to read from the slave the byte is 0xA3.

There are nine clocks for each byte transferred. During the last clock pulse the device sending data lets the data line float high. The receiver then pulls the line low before the clock goes low in order to acknowledge the byte was received. There is no standard as to what to do when data is not "ack'd." Some devices depend on this as a way for the master to stop the transmission. Sometimes the ack is only checked on the byte after a start condition to make sure the device exists on the bus and is healthy.

Figure 21.10 shows what a read from address 3 in our 2401 serial EEPROM looks like. First the master sends the address/direction and then the address he wants to read from the device. Then the direction is switched to read data from the slave and 1 byte is read.

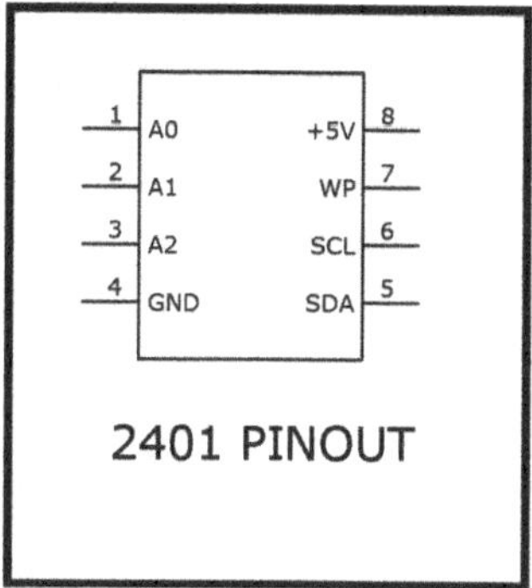

Figure 21.9: 2401 serial EEPROM pin-out.

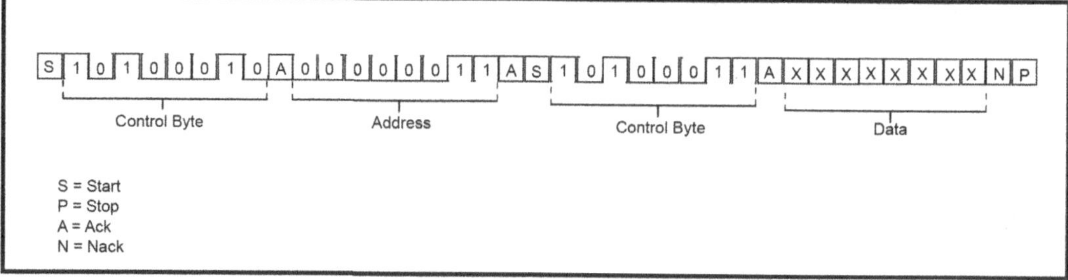

Figure 21.10: I²C bus EEPROM read timing diagram.

The source code looks like this:

```
#use i2c(master, sda=PIN_B1, scl=PIN_B2)
...
i2c_start();
i2c_write(0xA0);
i2c_write(3);        // Address in the EE we want to read
i2c_start();
i2c_write(0xA1);     // Change direction
data=i2c_read(0);
i2c_stop();
```

The way this serial EE works is it keeps sending bytes until you do not ack one. The **i2c_read()** by default acks bytes but if you pass a 0 argument it will not ack the byte.

Like the SPI, some PIC® devices have hardware I²C. More have slave support than master support. To use the hardware, as you might guess, it looks like this:

```
#use i2c(master, I2C1 )
```

You can also set the speed using fast and slow, like this:

```
#use i2c(master, sda=PIN_B1, scl=PIN_B2, fast)
```

The above code did not check to see if the slave was there. A check for the slave ack would look like this:

```
i2c_start();
if(i2c_write((0xA0)!=0)
   cout << "Error on I2C bus" << endl;
else {
   i2c_write(3);
   i2c_start();
   i2c_write(0xA1);
   data=i2c_read(0);
   i2c_stop();
}
```

The **i2c_write()** returns a 0 for ack and 1 for no ack.

Multi-master

The I²C bus can have multiple devices that are the master. There can only be one master communicating on the bus at a time. The master is usually accustomed to grabbing the bus whenever he wants. In a multi-master situation it must first check to see if the bus is in use. In this case **i2c_write()** will return a 2. The code for multi-master looks like this:

```
#use i2c(multi_master, sda=PIN_B1, scl=PIN_B2)
...
i2c_start();
while(i2c_write(0xA0)==2)) ;
i2c_write(3);
i2c_start();
i2c_write((0xA1);
data=i2c_read(0);
i2c_stop();
```

Table 21.1 Special I^2C addresses.

00000000	This is broadcast address. It is used if the master wants to send information to all the slaves at once
00000001	This address is used by a not-so-intelligent master device. For example, a keypad who does not know the slave address of the slave it needs to talk to (when a button is pressed). There is always another byte following this one that has a unique address. In this case it is a master address that a slave can recognize as someone it want to receive data from. Despite the LSB 1 the master writes data with this special address
0000001x	This address is used for another kind of bus called CBUS. CBUS uses another wire but the SCL and SDA can be shared with I^2C devices. When this address is seen by a CBUS device it starts using the third wire. I^2C simply ignores activity until the stop condition. CBUS is not too popular so it is not covered in this chapter
0000010x	Called a START byte, this special address is used to kill time before another start and the real address. This is used if the slave is slow and does not have interrupts. It is used to alert the slave that is polling the bus that something good about to come
0000011x 00001xxx 11111xxx	Reserved and should not be used. These patterns may be defined in the future as some special purpose
11110xxx	Used for 10-bit addresses, see next section

Special Addresses

The I^2C bus has defined some addresses to have special meaning. Table 21.1 identifies the special addresses. Shown in the first column is the full 8 bits in the first byte after a start condition.

10-Bit Addresses

The I^2C bus slaves can have a 10-bit address, as opposed to the 7-bit address described thus far.

It works by using 2 bytes for the address/direction instead of one. The most significant 5 bits of the first byte are 11110 and this indicates a second address byte is coming. The low 2 bits of the 7-bit address byte are then the most significant 2 bits of the new 10-bit address. For example, a device with an address of 0b0101010101 would be sent like this (in write mode):

```
i2c_start();
i2c_write(0b11110010);
i2c_write(0b01010101);
```

It is perfectly normal to have both 7-bit address devices and 10-bit address devices on the same bus. A 7-bit device will ignore all data until a stop condition if the first address byte does not match. No 7-bit address is allowed that starts 11110, so there is no problem.

Slave I²C

Similar to SPI, sometimes a processor wants to be a slave. You should always use hardware I²C for a slave I²C because firmware cannot detect the start and stop conditions reliably. For PIC16/18 parts the same **int_ssp** is used for the I²C interrupt. You cannot do hardware I²C and SPI at the same time with the same hardware unit. Some chips have up to four units. The interrupt code for a 2401 emulator would look like this:

```
#use i2c(slave, address=0x80, sda=PIN_B1, scl=PIN_B2)
...
#INT_SSP
void ssp_interrupt ()  {
   unsigned int8 incoming, state;

   state = i2c_isr_state();
   if(state <= 0x80)  {                     //Master is sending data
      if(state == 0x80)
         incoming = i2c_read(2);
      else
         incoming = i2c_read();

      if(state == 1)                        //First received byte is address
         address = incoming;
      else if(state >= 2 && state != 0x80)  //Received byte is data
         buffer[address++] = incoming;
   }
      if(state >= 0x80)                     //Master is requesting data
         i2c_write(buffer[address++]);
}
```

The magic in this function is the **i2c_isr_state()** function. It determines what state the bus is in. The return values are:

0x80	master just sent address and requests read mode
0x00	master just sent address and requests write mode
0x01-0x7F	master just sent byte n of data
0x81-0xFF	master is requesting byte n-0x80 of data

The `i2c_read()` argument is a 0 for no ack, 1 for ack, and a 2, as used above, means to stretch the clock. This is used to hold the clock until we do the write (a few lines below it).

Note that this function not only handles reads from the emulated EE but also deals with writes to the emulated EE.

The address in the `#use i2c` is the 7-bit address shifted up 1 bit. This is frequently the way it appears in data sheets.

SMBus

The system management bus is a specification that is built upon the I²C interface with standardized packet and command formats defined at a higher level. It is now used heavily in the smart battery industry. If you look at a cell phone or laptop battery you may see four terminals. Two go to the battery (+ and −) and the other two are probably SMBus (SCL and SDA).

Summary

- SPI is a popular hardware protocol for communicating with simple electronic devices.
- SPI has a data out (SO), data in (SI), and clock signal in addition to the chip select.
- There are four modes of SPI data transfer depending on the clock idle state and the edge the data is sampled on.
- Multiple SPI devices can be on the same bus as long as each has its own chip select.
- Some SPI devices use an additional signal to indicate framing.
- The I²C protocol is always two wires and works based on each device having a unique address.
- I²C always transfers data in multiples of a byte and each byte can be acknowledged.
- Standard I²C speeds are 100K and 400K bits per second and if needed the slave can hold up the master if it is busy.
- Transactions begin with a special start condition and then after some number of data bytes a stop condition.
- A start condition can happen again before the stop condition to change data transfer direction.
- The first byte after a start condition specifies the slave address and data transfer direction.
- The I²C address is usually 7 bits; however, a 10-bit address is defined.
- The I²C protocol allows for multiple masters to be on the same bus with each master detecting the others and waiting until the bus is free before using it.
- Some I²C addresses are reserved for a special purpose.

Exercise 21-1

Objective: Gain a working knowledge of using I^2C and SPI busses with C as the programming language.

Requires: E3 module, USB cable, PC, 25LC040A-IP, TC74A-05.0AT, breadboard, eight jumper wires.

Steps/Technical Procedure	
1. Write a program that accepts a command of R (read) or W (write). For write, accept lines of text entered until a line is entered with no characters on it. Each line should be saved in an external SPI serial EEPROM starting at address 0 of the device. The R command should play back the entire message to the screen. The serial EEPROM pin-out is as follows. The ~name is used instead of name to indicate active low on schematic tools that do not support overstrike. The chip has a simple 3-byte command to read data and a 3-byte command to write data to the chip. Addresses in the chip are from 0-0x1FF. Use the chip data sheet to get the details. Search mouser.com or digikey.com for the part number and click "Data Sheet" to get the data sheet. Section 2.0 has the full protocol details. **HINTS:** • Do not forget the CR/LF at the end of each line in the EEPROM and some kind of message termination character so you know when to stop reading. • The WP and HOLD pins don't need to be connected to the E3 board. They should be connected to Vdd or Vss, however. • Make sure you have a check to terminate input if you run out of space in the EEPROM.	
2. Write a program that displays the temperature on the screen in Fahrenheit every 5 seconds using an external I^2C temperature sensor. The temperature sensor pin-out is as follows. The chip has a simple read command to read data and there is no need to write to this part. Use the chip data sheet to get the details. Search mouser.com or digikey.com for the part number and click "Data Sheet" to get the data sheet. Section 2.0 has the full protocol details. **HINTS:** • The chip output is a signed byte and in Celsius. You will need to make a conversion to Fahrenheit.	

Quiz

(1) What key characteristic makes the SPI bus synchronous as opposed to asynchronous?
  (a) The S in SPI
  (b) The bidirectional data transfer over two wires
  (c) The chip select pin on each device
  (d) The clock signal
  (e) The firmware algorithm

(2) What is the absolute minimum number of signals you might see between an SPI master and slave?
  (a) 1
  (b) 2
  (c) 3
  (d) 4
  (e) 5

(3) How many PIC pins are required for a five-device SPI bus that requires full concurrent bi-directional communication?
  (a) 2
  (b) 3
  (c) 4
  (d) 6
  (e) 8

(4) For the following slave code, what should the master code look like?

```
#int_ssp
void spi_slave_isr(void) {
    static int16 data;
    static int8 cycle;

    switch(cycle++) {
        0 : data=spi_xfern();  break;
        1 : data=data | (int16)spi_xfern() <<8;
             spi_prewrite((data&0xFF)^0x12); break;
        2 : spi_prewrite((data>>8)^0x12); break;
        3 : cycle=0;
    }
}
```

(a) `datain=spi_xfer(dataout & 0xFF) | spi_xfer(dataout>>8)`
(b) `spi_xfer(dataout); datain=spi_xfer(0)`
(c) `datain=spi_xfer(dataout)`
(d) `datain=spi_xfer(dataout) & spi_xfer(0)`
(e) None of the above will work

(5) For a 500-kHz clock, how long will it take to send out 10 bytes to a slave device?
(a) 10 us
(b) 20 us
(c) 80 us
(d) 160 us
(e) 200 us

(6) In the 7-bit address mode, how many I²C devices can be on the same bus?
(a) 1
(b) 97
(c) 112
(d) 127
(e) 255

(7) In terms of the SPI modes, what mode would the I²C data transfer be?
(a) 0
(b) 1
(c) 2
(d) 3
(e) 4
(f) None

(8) Of the following, what is the false statement for I²C?
(a) An address/direction byte ALWAYS follows a start condition
(b) The number of start conditions on a bus should equal the stop conditions
(c) All data bytes have an ack bit
(d) 7-bit address devices and 10-bit address devices can be on the same bus
(e) Address/direction bytes have an ack bit

(9) With a 100-kHz I²C clock, how long will it take to transfer 100 data bytes between the start condition and stop condition?
(a) 100 us
(b) 4 ms
(c) 8 ms
(d) 9 ms
(e) 11 ms

(10) An LCD controller has a protocol that indicates the master must first send a byte address on the display line that data should be written to and then the data can be sent. What is wrong with the following code to write ABC to the display?

```
i2c_start();
i2c_write(0x90);
i2c_write(0);
// set display address to the beginning
i2c_start();
i2c_write('A');
i2c_write('B');
i2c_write('C');
i2c_stop();
```

(a) Missing an address write after the second start
(b) Extra start that should not be there
(c) Can't send characters in **i2c_write**, must convert to bytes
(d) Need a **0x91** not **0x90** in the first write
(e) Nothing is wrong with the code

CHAPTER 22

External Serial Busses

Perhaps the most well-known external serial bus is the USB bus. The same things that make this such an easy interface for the user to use make it complex for the programmer. There is a complex multi-step process to link two devices via the USB bus. This is beyond the scope of this book.

RS-232 is also a very popular protocol, and provides a standard point-to-point communication for unit-to-unit communication. It is a bidirectional link where each device has a dedicated transmit signal. This is called full duplex. Serial peripheral interface (SPI) is also (or can be) full duplex, whereas I^2C is always half duplex. RS-232 can cleanly communicate through 50 feet of cable, whereas SPI and I^2C are more designed for on-board communication. RS-422 is a variation of the physical protocol that vastly extends the distance using an extra two wires. RS-485 is a variation of RS-232 that allows many devices to hang on the same bus (multi-drop). RS-232-like communication is also used for short distance processor-to-processor communication without the bus drivers usually needed for out-of-unit communication.

RS-232

The SPI and I^2C are synchronous busses. The clock can be sloppy, it can pause for a while and then start up and it all works because data is transferred on one of the edges. RS-232 uses an asynchronous protocol. The clock is not shared between nodes. The way it works is the signal sits at the 1 state (known as the mark state). To send data, first a start bit of 0 is sent (called a space) and then data bits are sent followed by a stop bit. The stop bit is also a 1 state. The receiver waits for the signal to drop from 1 to 0 and then it starts a timer. It knows there will be a new bit at a fixed time after the time that the line first changed and then another bit the same time thereafter. The sender and receiver must have already agreed on a bit time (baud rate) and their clocks must be at least 3% accurate for this to work. Data is sent LSB first. See Figure 22.1.

The stop bit guarantees the line goes back to 1 so the receiver can properly detect the next start bit. Every byte gets a start and stop bit. Some protocols specify 2 stop bits to give older equipment enough time to digest the data. The number of data bits can vary but usually it will be 8 bits, with some protocols using 7 or 9 bits.

Embedded C Programming. http://dx.doi.org/10.1016/B978-0-12-801314-4.00022-3

0 1 2 3 4 5 6 7

Start Bits Stop

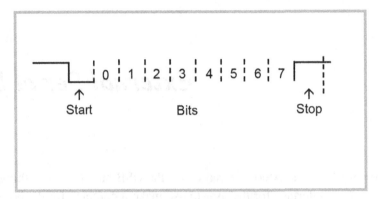

Figure 22.1: 8-bit data asynchronous timing diagram.

Table 22.1 Example of bytes showing parity.

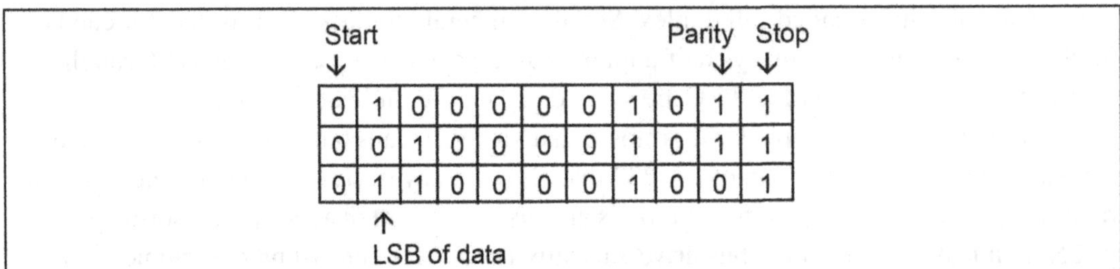

Start	LSB of data								Parity	Stop
0	1	0	0	0	0	0	1	0	1	1
0	0	1	0	0	0	0	1	0	1	1
0	1	1	0	0	0	0	1	0	0	1

Sometimes a parity bit is used to detect errors in transmissions. Even parity means the parity bit is 1 if there is an even number of 1s in the data. Odd parity means the parity bit is 1 if there is an odd number of 1 bits in the data. The parity bit will appear just before the stop bit. Table 22.1 shows some examples of even-parity data for 'A' to 'C' with the start and stop bits.

For hardware, the RS-232 standard indicates the voltage levels need to be from −3 to −25 V to indicate a mark (1) and from +3 to +25 V for a space (0). The voltage differential is what allows this to work to 50 feet. In reality it will work much farther. These voltages are not PIC®-friendly voltages so a level converter is required. We have one signal for each direction of communication and a common ground (intended for off-board). Figure 22.2 shows how this looks.

If the communication is onboard such as would be between processors, then the level converters are skipped; they already have a common ground. This is sometimes called TTL RS-232. See Figure 22.3.

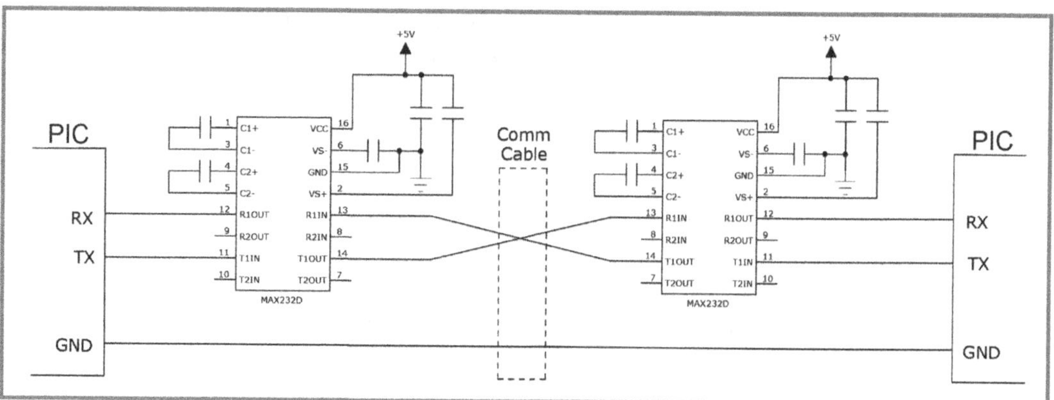

Figure 22.2: Complete PIC® MCU to PIC® MCU RS-232 connection diagram.

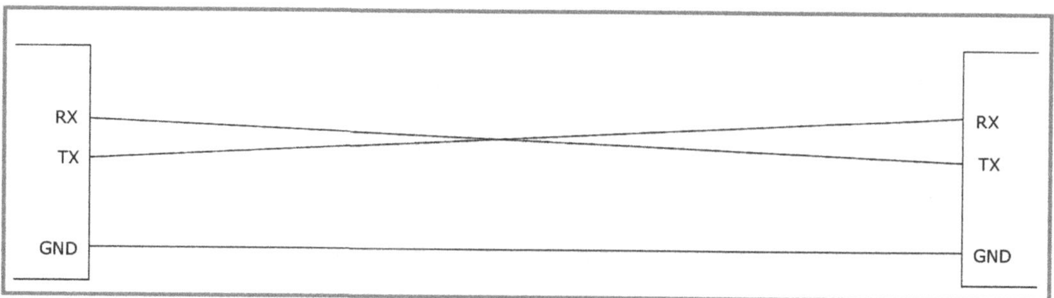

Figure 22.3: PIC® MCU to PIC® MCU connection diagram using TTL levels.

Source Code

For the E3 board we have already used **getc()** and **putc()** for the USB-to-PC communication. This could also be set up for RS-232, but instead we will define a stream name and use **fgetc()** and **fputc()**. Here is a simple program:

```
#use rs232(baud=9600, xmit=PIN_B0, rcx=PIN_B1, stream=SERIAL)
void main(void) {
   char c;
   fprintf( SERIAL, "Start Typing, echo back will be successive chars\r\n");
   while(TRUE) {
      c = fgetc(SERIAL);
      if(c>' ')
         c+=1;
      fputc(c,SERIAL);
   }
}
```

The stream name is used in each I/O call. We can also use **cin/cout**-like communication but we now replace the **cin** and **cout** with **SERIAL**, like this:

```
void main(void) {
   char s[80];
   SERIAL << "Type lines of text" << endl;
   while(TRUE) {
      s << strspace << SERIAL;
      SERIAL << "You typed: " << s << endl;
   }
}
```

In addition to the **cin/cout** statements you are now very familiar with, you can use the following functions with an RS-232 stream:

putc() / fputc()	send a single character
getc() / fgetc()	read a single character
puts() / fputs()	send a string and append a /r/n
gets() / fgets()	read a string terminated with a /r
printf() / fprintf()	formatted print
kbhit()	check to see if a character has come in

For **printf** the full set of format specifiers is as shown in Table 22.2.

Table 22.2 printf() specifier list.

%c	Character
%s	String or character
%u	Unsigned int
%d	Signed int
%Lu	Long unsigned int
%Ld	Long signed int
%x	Hex int (lowercase)
%X	Hex int (uppercase)
%Lx	Hex long int (lowercase)
%LX	Hex long int (uppercase)
%f	Float with truncated decimal*
%g	Float with rounded decimal*
%e	Float in exponential format*
%w	Unsigned int with decimal place inserted*

*Specifies two numbers for n. The first is a total field width. The second is the desired number of decimal places.

UART

A hardware peripheral that accepts and transmits RS-232-like data is called a universal asynchronous receiver/transmitter, or UART for short. Sometimes it will be referred to as a USART. The S is for synchronous because RS-232 does define a synchronous-type communication as well as asynchronous. In that case there are an extra two wires for a clock in each direction. Once popular, it is almost never used any more so we are not going to waste any more space on synchronous RS-232.

Most PIC® processors have a UART in them and some have as many as four. The compiler interprets the **#use rs232** directive in one of two ways depending on the pins specified for the communication. If the pins are connected to a UART then code to work with the UART is generated. If the pins are not connected to a UART then the compiler generates code to emulate the UART operation with just firmware. In addition, instead of specifying pins, you can just put UART1 or UART2 to use the built-in UART on whatever pins it is connected to.

There are some advantages to using the UART. While data is coming in (a byte) the processor can be doing other things. Without a UART the processor is stuck in **getc()** until a character has been received or in **putc()** until it has gone out. In addition, **kbhit()** only returns true if you have started receiving the start bit and if you don't call **getc()** quickly you lose the character. Although you can set up interrupts if you use an interrupt pin without a UART, with a UART you can do interrupts at a higher baud rate. Finally you should be aware there is a three-character incoming byte buffer and two-character outgoing buffer when using the UART on most chips. This means if you do not call **getc()** fast enough you can wait up to almost three character times before losing data.

When using the UART you can change the baud rate at run time using a call like this:

```
setup_uart( 4800 );
```

The initial speed is specified in the **#use rs232** and then you can change it at any time at run time. This is sometimes done when a switch setting sets the baud rate and the switch is read when the program starts up. There are also some more complex protocols where one of the devices on the bus makes a request to change the baud rate. This happens most often when the power-up state is at a slow baud rate, and then after establishing communication a device will indicate how fast it can run and if both devices agree they change the baud to run faster. Care must be taken not to change the baud rate until the communication lines are idle. For example, a common mistake is to do this:

```
cout << "Baud changing to 19200" << endl;
setup_uart( 19200 );
```

Because of buffering, the **\r\n** (and maybe the last "0") did not yet get out and the baud rate is changed.

Some UARTs have additional features and these are invoked with **setup_uart()**. For example, some parts can send what is called a break condition. This essentially holds the transmit line low for a half second. It is used in some protocols as a kind of reset signal. Another feature some parts have is to detect the baud rate of the other device when the device sends a pre-defined character at a specific time.

Incoming Data Interrupts

Interrupts are very easy to use with the UART. You get an interrupt each time a byte has come in. Here is an example that collects a line's worth of data and the main program prints it out.

```
#define LINE_SIZE 60
char  line[LINE_SIZE];
int8 next_in = 0;
int1 line_in = FALSE;

#int_rda
void serial_isr() {
   char c;

   c=getc();
   if(line_in)
      return;

   if(c=='\r') {
       line[next_in]=0;
       line_in=TRUE;
   } else
        line[next_in++]=c;
}

void main(void) {
   enable_interrupts(INT_RDA);
   enable_interrupts(GLOBAL);

   while(TRUE) {
      while(!line_in) ;
      cout << "Data=> " << line << endl;
      line_in = FALSE;
   }
}
```

The **INT_RDA** stands for receive data available. Many UARTs use this terminology for incoming data.

The above code is nicely synchronized between the interrupt service routine (ISR) and **main()** with the **line_in** variable. The only problem might be that data keeps coming in after the first line and **main()** has not yet printed the data and cleared the flag. In that case data is lost at the start of the next line. If we get rid of the **if(line_in) return;** in the ISR then new data would overwrite old data. One solution is to use a double buffer: a **line1** and **line2**, and the **line_in** variable indicates which line is full. This still could have trouble if **main()** is busy and does not read the data fast enough. A common solution to this problem is to use a ring buffer.

For a ring buffer, data always just keeps coming into the buffer and when the end is reached it wraps around to the start of the buffer. The plan is for the **main()** program to have read that data by the time it wraps around. If it can't, you simply make the buffer larger. There are two pointers, one to the next position to write to (**next_in**) and one for the next position to read from (**next_out**). For our program, since it is record based (lines), we also will have a line counter. Here is the program:

```
#define BUFFER_SIZE 256
char  buffer[BUFFER_SIZE];
char  next_in = 0;
char  next_out = 0;
int8  lines_in = 0;

#int_rda
void serial_isr() {
   int t;
   buffer[next_in]=getc();
   if(buffer[next_in]='\r')
      lines_in++;
   t=next_in;
   next_in=(next_in+1) % BUFFER_SIZE;
   if(next_in==next_out)
     next_in=t;              // Buffer full !!
}

void main() {
   char c;
   enable_interrupts(INT_RDA);
   enable_interrupts(GLOBAL);

   while(TRUE) {
      while(lines_in==0) ;
      printf("Data => ");
```

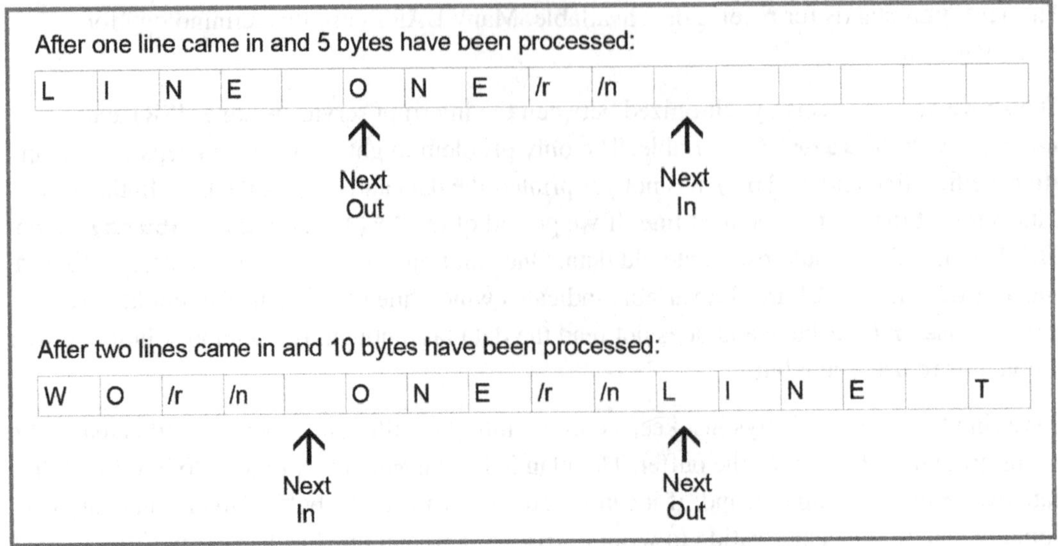

Figure 22.4: Illustration of a ring buffer.

```
        do {
            c=buffer[next_out];
            next_out=(next_out+1) % BUFFER_SIZE;
            putc(c);
        } while(c!='\r');
        putc('\n');
        lines_in--;
    }
}
```

Figure 22.4 is a diagram of the buffer and pointers after the first line comes in, as the second line is coming in, and after the buffer wraps around. You can see the **next_out** pointer is simply chasing the **next_in** pointer.

Outgoing Data Interrupts

UARTs also provide interrupts for outgoing data. At 9600 baud it takes a millisecond for each character to go out. A 50-character line then takes 50 ms. This is a lot of time for some applications to spend stuck in printf. The alternative is to stuff the 50 characters into a buffer and then allow an interrupt process to send the data out. The code looks like this:

```
#define T_BUFFER_SIZE 64
byte t_buffer[T_BUFFER_SIZE];
byte t_next_in = 0;
byte t_next_out = 0;

#int_tbe
void serial_isr() {

   if(t_next_in!=t_next_out) {
      putc(t_buffer[t_next_out]);
      t_next_out=(t_next_out+1) % T_BUFFER_SIZE;
   } else
         disable_interrupts(int_tbe);
}

void bputc(char c) {
   int1 restart;
   int8  ni;

   restart=t_next_in==t_next_out;
   t_buffer[t_next_in]=c;
   ni=(t_next_in+1) % T_BUFFER_SIZE;
   while(ni==t_next_out);
   t_next_in=ni;
   if(restart)
      enable_interrupts(int_tbe);
}

void main() {
   enable_interrupts(GLOBAL);
   bputc << "This is a test output line" << endl;
   while(TRUE) ;
}
```

TBE stands for transmit buffer empty. This interrupt happens anytime the UART is ready for another byte of data. Because the only way to clear that interrupt is to put data into the UART buffer, we need to manipulate the INT_TBE enable flag more than we usually do for interrupts. In this example we set the enable flag whenever we put data into the RAM buffer. It is cleared after we take the last data byte out of the buffer. That clearing prevents any further interrupts until new data is put into the RAM buffer. It should also be clear this code also uses a ring buffer. Streaming is used to send the character data to a function that inserts the data into a buffer.

Modem Control Signals

In the era when the RS-232 standard was developed, modems were the method to connect terminals to computers. To help the terminal and modem out, some additional signals were

Table 22.3 RS-232 modem control signals.

RI	Ring indicator	DCE to DTE
DCD	Data carrier detect	DCE to DTE
DTR	Data terminal ready	DTE to DCE
DSR	Data set ready	DCE to DTE
RTS	Request to send	DTE to DCE
CTS	Clear to send	DCE to DTE

defined by the RS-232 standard (see Table 22.3). For example, the RI signal is used to tell if there is an incoming call on the modem phone line (ring indicator). The standard defined two types of devices on the bus. One side is a DTE (data terminal equipment) and the other side is a DCE (data communications equipment). It is clear the RX (receive) pin on the DTE connects to the TX pin on the DCE and viceversa.

Often these signals are used for purposes other than the original intent. For example, an RS-232 printer might use RI to indicate it is out of paper. Frequently the modem control signals are not used at all.

Hardware Flow Control

One popular use of the modem control signals is for hardware flow control. This is used to stop a device from sending data when the receiver is not able to accept it. This may be because the receiver buffer is full. For example, to send a byte of data the code may look like this:

```
while(input(RTS_PIN));
putc(c);
```

This causes the processor to wait for RTS to be low before sending a byte. The **#use rs232** directive allows you to specify the modem control pins so you don't have to have this logic in your code. Be aware that a call to **putc()** could hang until the receiving device is ready when you use modem control signals for hardware flow control.

Software Flow Control

A similar flow control can be done without extra wires. One popular method is called XON/XOFF. These are two ASCII-defined characters. XON is 0x11 and XOFF is 0x13. When the receiving buffer is nearly full, the receiver sends an XOFF and the sender then holds and waits for an XON before sending more data. This only works if a 0x11 and 0x13 will never appear in the normal data stream. This would be the case for standard text (all characters are 0x20 and up).

Example code for a receiver that reads a line of data and then processes the data might look like this:

```
#define XON  0x11
#define XOFF 0x13

while(TRUE)  {
   putc(XON);
   cin >> string;
   putc(XOFF);
   process_command(string);
}
```

On the other side the code might look like this:

```
void send_byte( int8 c ) {
   int8  in=0;

   while(kbhit()) {
in=getc();
      if(in==XOFF)
         while(in!=XON)
in=getc();
   }
   putc(c);
}
```

Protocol

Since the standard does not provide any guidance for a higher level protocol, there are many protocols for RS-232. Usually when communicating with a device, the device will have a document to describe how to communicate with it. Included will be baud rate, number of bits, and sometimes the method of flow control. Beyond that there will be the format of the byte data stream in both directions. To get a feel for these protocols we will briefly describe a well-used protocol (from NEMA) used by most serial GPS devices. All messages in both directions start with a **$** and end with a **\r\n**. This means a receiver can discard incoming data bytes until a **$** is seen. Table 22.4 shows the format of a specific command to the GPS.

The GLL data (get longitude/latitude) from the GPS is shown in Table 22.5.

A sample command would then be:

```
$PSRF103,01,01,00,01*26\r\n
```

Table 22.4 NEMA GPS query/rate control message.

$	Start of message
PSRF103,	Indicates message type, PSRF103 means this message controls data the GPS unit sends out
n,	Type of message wanted, 1 means GLL (will cover below)
n,	1 means send once, 0 means send at some rate
n,	Number of seconds between transmissions when above is 0
n,	0 means don't send checksum, 1 means send checksum
*n	n is the byte checksum in hex of all data prior to this, including the $, not including the *, and is in hex.
/r/n	End of message

Table 22.5 NEMA get longitude/latitude response.

$	Start of message
GPGLL,	
ddmm.mmmm,	Latitude
n,	n is N for north or S for south
ddmm.mmmm,	Longitude
n,	n is W for west or E for east
hhmmss.fff,	Time in hours, minutes, seconds, and fractions of a second
n	N is A for valid data or V if this data is not valid. Not valid because of a poor signal
*n	n is the byte checksum in hex of all data prior to this, including the $, not including the *
\r\n	End of message

The response would be

```
$GPGLL,4259.3892,N,8815.1346,W,111304.174,A,A*93\r\n
```

RS-232 Future

RS-232 is becoming less popular and USB is becoming more popular. Many modern PCs don't even have an RS-232 port; that, however, does not mean the concepts in this chapter are outdated. One of the modes USB operates in is the CDC mode and that is designed to mimic an RS-232 port. That means for devices that were designed with RS-232, switching to CDC USB can be done without changing the code. The protocol remains the same. Likewise new designs are using CDC because the communications techniques are well understood by programmers.

RS-422

RS-422 is a variation of RS-232 where the only difference is the physical layer. With protocols, we refer to the physical layer as the electronic part of the protocol (the wires and level shifters). The software for RS-422 is the same. The physical layer uses a differential signal for each signal. This means two wires for each direction. When one wire is high the other is low (see Figure 22.5). Instead of looking for a specific voltage level, the receiver simply looks to see which wire is higher in voltage than the other to determine if it is a 1 or 0. This extends than range of the wire to around 4000 feet.

The firmware is all the same and it does not need to know if the level translators are RS-232 or RS-422.

RS-485

RS-485 is kind of an extension to RS-422 to make it multi-drop (as opposed to point-to-point). This means many devices can hang off the same bus. There is only one signal connection consisting of two wires differential just like RS-422. RS-485, unlike RS-232 and RS-422, is always half duplex and does not have separate signals for each direction. The RS-485 protocol does not deal with how to figure out who can send data on the bus. This is left to the programmer. The following is a protocol description that is very typical for RS-485 protocol designs:

- Word format is 9 data bits with 1 start bit and 1 stop bit.
- If the MSB is a 1 then the other 8 bits are a control byte. A simple control command might just be the address of the device we need to communicate with.
- If the MSB is a 0 then the other 8 bits are data bytes. The protocol may have a convention that says after all data is received for a device that device responds with any data it has for the master or just an acknowledgment of data received.

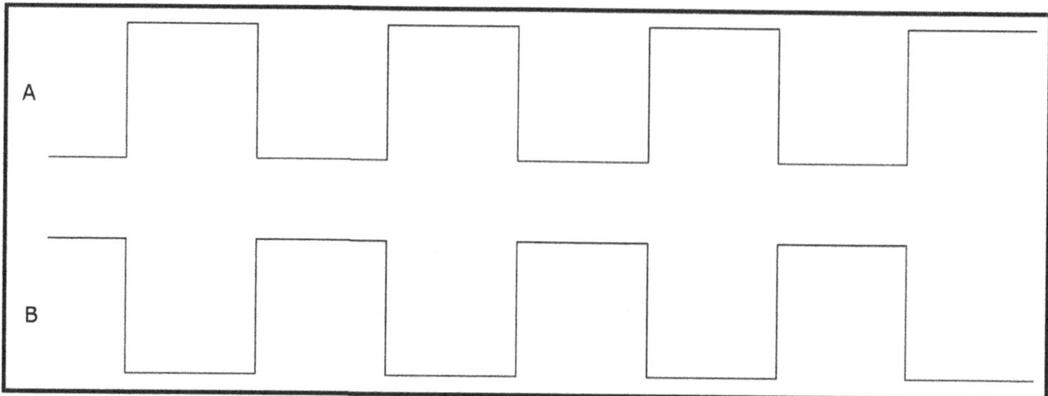

Figure 22.5: Differential data transmission diagram.

You can see there is a master device that controls the bus. The master will typically poll each device to transfer data to that device. No one except the master speaks on the bus without permission. All devices monitor the bus and if a control byte comes in without their address then all data is ignored until another control byte comes in.

There is one more twist that needs to be considered with RS-485. The way the level converters work is they need to be told, via a pin, if they should transmit or receive. If you control this pin manually be aware you must not switch from transmit back to receive until the UART transmit buffer is empty.

The compiler has built-in functions to deal with all this. Here is what the preprocessor directive looks like:

```
#use RS-232( baud=9600, bits=9, LONG_DATA,    \
   xmit=PIN_C6, rcv=PIN_C7, enable=PIN_C5)
```

The **bits=9** sets the number of data bits to 9. The **LONG_DATA** changes **getc()** and **putc()** to work with int16 instead of int8. The **enable=** is used to identify the R/T pin on the level converter. See Figure 22.6.

Documentation

When creating a new RS-232-like interface it is critical to first create an interface document fully describing the protocol. Don't forget how error situations are handled. What happens if the checksum does not match? What happens if you get a start of message but no terminator? How long do you wait? What is then done to re-synchronize communication? Including the version number of that document in the protocol is also a good idea. Perhaps in the introductory exchange at the start of establishing communication, so both devices agree they are using the right protocol.

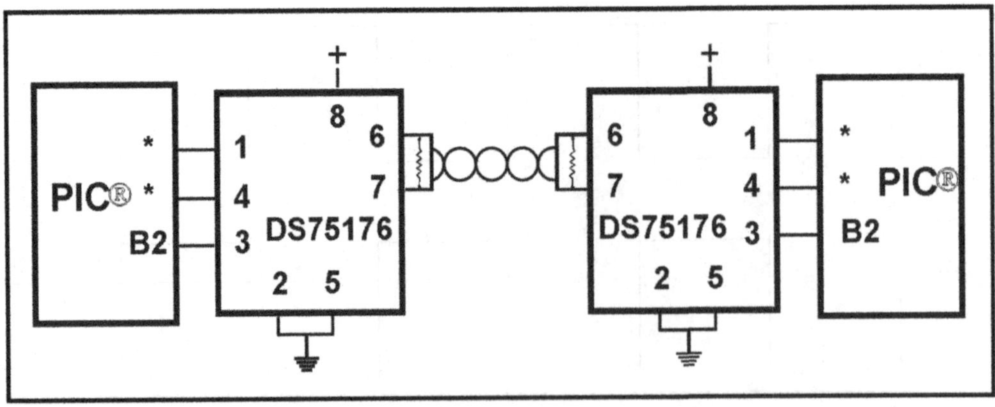

Figure 22.6: Complete PIC® MCU to PIC® MCU RS-485 connection diagram.

A lot can go wrong with these protocols. Think early about how the situations will be handled. Make sure all the programmers for all the devices have the protocol before coding begins. It will save a lot of time.

Summary

- RS-232 is an asynchronous serial protocol used to communicate to external devices using a voltage range of at least −3 to +3 V. Negative voltages are a 1, positive are 0.
- The RS-232 word format includes a 0 start bit, some number of data bits, optional parity bit, and a stop bit.
- TTL RS-232 is used for short range communication without level converters.
- Standard C functions (**`putc()`**, **`getc()`**, **`puts()`**, **`gets()`**, **`printf()`**, **`kbhit()`**) may be used to facilitate RS-232 communication with the **`#use rs232`** directive.
- Any pins may be used to RS-232; however, more features are available by using a UART and the UART pins.
- Incoming and outgoing data interrupts are available on PIC® devices with a UART.
- Modem control signals defined in the RS-232 standard may be used for hardware flow control, for purposes not related to the protocol, or not at all.
- Software flow control is relatively easy to implement if the protocol allows for some reserved symbols.
- RS-422 is a long distance version of RS-232 that uses more wires but no firmware changes.
- RS-485 is a multi-drop version of RS-232 where typically each device has its own address.

Exercise 22-1

Objective: Gain experience using RS-232-type communication and dealing with a command/response-type protocol.

Requires: E3 module, USB cable, PC, TTL RS-232, GPS unit.

Steps/Technical Procedure	Notes
1. Similar to Exercise 12-1, write a program that accepts commands and then acts on those commands. The program should display nice errors for improperly formatted commands. The commands to accept are: LED RED ON LED RED OFF LED GREEN ON LED GREEN OFF LED YELLOW ON LED YELLOW OFF	

(continued)

Steps/Technical Procedure	Notes
WRITE location data (location is a hex number, 0–7F, that writes data to an array) READ location (location is a hex number, 0–7F, that reads data from an array) POT (respond with voltage from the POT) location in RAM) BUTTONS (show state of each button) However: The above actions should be done on a second E3 board where a second program will reside with NO USB communication. The two boards should be linked by wires using TTL RS-232. Two people may work on this project, one for each program. A protocol interface document must first be written to document the protocol between the boards. It is assumed the full text as shown above is not going to be sent. The protocol should be designed with a checksum on every packet and should be designed to easily be expanded in the future. The E3 board has a UART connected to transmit pin B7 and receive B5. **HINTS:** • In addition to transmit and receive, it is a good idea to connect the grounds between the two boards.	
2. Write a program that connects to a TTL RS-232 GPS unit and displays every 5 seconds the latitude, longitude, and altitude. If the data is not valid, display "Poor Reception." NEMA 0183 is the GPS standard. Use Google to find a copy on the web. **HINTS:** • Make sure to use the default baud rate in the standard to first connect.	

Quiz

(1) If no data is being sent over a true RS-232 link, what are possible voltages for the RX and TX pins of a device on the bus?

(a) RX = 5V TX = 5V

(b) RX = −5V TX = −5V

(c) RX = 5V TX = −5V
(d) RX = −5V TX = 5V
(e) RX = 0V TX = 0V

(2) What is the maximum number of devices that can be on an RS-232 bus?
(a) 1
(b) 2
(c) 8
(d) 127
(e) There is no limit

(3) How much time will it take to send 100 8-bit bytes on a 9600 baud RS-232 link?
(a) 83 ms
(b) 104 ms
(c) 125 ms
(d) 135 ms
(e) 270 ms

(4) What happens on RS-232 if two devices on the bus attempt to start sending data at exactly the same time?
(a) If that is possible, modem control signals need to be used to prevent it
(b) The collision must be detected by both senders and they must stop and wait a random amount of time before retrying
(c) The data is corrupted and will be detected by a bad checksum
(d) This is an indication of a poorly designed protocol and the results will be unknown
(e) Nothing bad will happen

(5) The following code is to receive a message over RS-232. Which of the following statements is true concerning this protocol?

```
while(getc()!=1) ;
buffer[0]=getc();
for(i=1,cs=0;i<=bufer[0];i++)
    cs+=buffer[1]=getc();
if(getc()!=cs)
    cout << "Error" << endl;
```

(a) Messages start with a 1 and end with a 0 and there is a checksum of all data
(b) Messages start with two 1s, then terminate when an incoming byte matches the checksum up to that point
(c) Messages start with a 1 then the command is sent twice, followed by data and a checksum

(d) Messages start with a 0x01, the message length follows the start and it ends with a checksum
(e) There is an error in this code and it will not do anything useful

(6) Which of the following tasks can only be done with a hardware UART?
(a) Use of **`kbhit()`**
(b) Interrupt on receive data
(c) Not stall the program if data is not yet available
(d) Fullduplex
(e) All of the above

(7) The primary reason for using RS-422 over RS-232 is what?
(a) Fewer wires
(b) Longer distance
(c) No level converters needed
(d) Newer standard
(e) All of the above

(8) Which of the following methods can be used to prevent multiple transmitters on the RS-485 bus at the same time?
(a) Each device has a dedicated time slot to transmit in
(b) Each device transmits only when commanded by a master
(c) Each device transmits only after a predefined other device transmits
(d) Each device transmits only when a separate signal pin input goes high
(e) All of the above

(9) Why does RS-485 frequently use a 9-bit data size?
(a) So a device does not confuse normal data for an address
(b) To correct errors in transmission
(c) To allow for 512 devices on the bus
(d) To get out more data in each transmission word
(e) All of the above

(10) For the following applications and bus picks, which one is not the best pick of the three we have studied?
(a) Industrial cabinet with several large processor boards within, using RS-422 to communicate between them
(b) Dual processors on a PCB, to exchange data between them a TTL RS-232 is used
(c) Thermostats in a building, all on a single RS-485 bus to report to a control unit
(d) Security panel in one building connected to a panel in another building in an industrial complex uses RS-422
(e) An appliance allows for a remote monitor to be connected using an RS-232 interface

```
    self_test_io();
}
#task(rate=100ms,max=10ms)
void task_three ( ) {
    read_and_filter_analog_data();
    check_button_presses();
    check_for_errors();
}
void main ( void ) {
    rtos_run ( );
}
```

First, the **#use rtos** is what brings in the RTOS library. Specified is the timer the RTOS can use and it indicates the smallest time increment that will be used. This is the shortest time between executions of a single task.

Each task function is tagged with a **#task**. The **#task** specifies the rate the function should be called at and the budgeted time for the task.

The first task is easy, it toggles a pin every second. The second task has calls in to **rtos_yield()**. This causes the task to give up its time slot and returns to the RTOS. You do not see it, but all task functions are in a **while(TRUE)** loop with a yield at the bottom. In **task_two()** we do some yields in the code. This may allow the task to meet its budget for time. It is sometimes done in a conditional statement based on whether the task has something to do.

The **rtos_run()** call in main starts the dispatcher.

One significant limitation in this chip-based dispatcher is any RTOS call that might result in the task losing its time slot must be made from the task function, not a function it calls.

Semaphores

A semaphore is a basic building block in operating systems. When we covered interrupts we used disable/enable interrupts to control access to shared data items. With multitasking we use semaphores to control access to shared data or any shared resource. On the surface a semaphore is a simple variable. We start the value out at the number of tasks that may use a resource at the same time. Consider a printer that might be used by multiple tasks (see Table 23.1).

Table 23.1 Example semaphore increment/decrement sequence.

Initialization	Set to 1 user of the printer	1
Task A wants to use the printer	Semaphore is >0 so it is decremented (called a wait)	0
Task A starts using the printer		0
Task B wants to use the printer	Semaphore is 0 so task B is suspended (in the wait)	0
Task A is finished	Semaphore is incremented (called a signal)	1
	Because the semaphore went positive task B is rescheduled to run and semaphore is decremented	0
Task B starts using the printer		0

In C it looks like this:

```
int8 sem_printer = 1;
#task(rate=2000ms,max=100ms)
void mytask ( ) {
   if( data_pending ) {
      rtos_wait( sem_printer );
      print_pending_data();
      rtos_signal( sem_printer );
   }
}
```

`rtos_wait()` waits for the semaphore to be greater than zero and then decrements it. **`rtos_signal`** increments the semaphore and restarts the task waiting the longest, if there is one.

Semaphores can be used in many ways in a multitasking environment. Consider this code:

```
int8  sem_rda = 0;
#int_rda
void rda_isr(void) {
   buffer[buffer_ptr++]=getc();
   rtos_signal( sem_rda );
}
#task(rate=2000ms,max=100ms)
void mytask ( ) {
   rtos_wait( sem_rda );
   process_data();
   rtos_signal( sem_rda );
}
```

The task starts and hangs right away on the semaphore. After the interrupt comes in the semaphore is incremented and the task now runs. It will run once for every byte that comes in through the receive data available (RDA) interrupt. If many bytes come in while other tasks are running, the semaphore keeps incrementing, so it is in effect the number of bytes that have come in. This is also the number of times wait can be called without hanging.

Message Passing

In a multitasking system we want to minimize the use of shared data. It becomes a real mess trying to ensure tasks are accessing global data in a safe manner. A basic OS tool to help in this mission is messages. Consider this example:

> We have a task that handles all the indicator lights on unit. This is the LIGHTS task. When someone wants to modify the lights it sends a message to the LIGHTS task. No other task touches the lights. The LIGHTS task will read a message and, based on the request and taking into consideration other requests, will decide what to do with the lights. For example, one task may send a message to turn on the error light. Then another task does the same. Finally the first task says to turn off the error light. The LIGHTS task knows to keep it on until it gets an all clear message from the second task.

This is a very orderly way for tasks to operate. Here is the code for the main LIGHTS task:

```
#task(rate=2000ms,max=100ms)
void lights ( ) {
   int8 cmd;
   cmd = rtos_msg_read ( );
   process_cmd( cmd );
}
```

rtos_msg_read() suspends the task until a message is in the queue. A message is sent like this:

```
rtos_msg_send(lights, cmd_request);
```

await()

The following is a common code block when using an RTOS:

```
while( !expr )
   rtos_yield();
```

It waits for some condition to be true and only hangs the task, not the processor. The RTOS has a built-in function to do the same thing. It looks like this:

```
rtos_await( expr );
```

The function allows other tasks to run until the expression is true.

Task Management

In addition to waiting for an event such as a semaphore, task execution can be manually blocked. A task itself can do this or another task can do it. It could also happen in an interrupt function or in **main()** before **rtos_run()** is called. There may, for example, be tasks that do not normally run until something happens. The following two functions are used for task control:

```
rtos_enable( task_name );
rtos_disable( task_name );
```

The deterministic dispatcher OS relies on each task staying within the allotted time. If a task takes too much time the other tasks shift off schedule. Sometimes you will want a task in charge of keeping an eye on the other tasks. One function that helps in this is **rtos_overrun()**. It can be used to identify a task that overran its time. It works like this:

```
if(rtos_overrun(task_A))
   cout << " Task A overrun" << endl;
```

The RTOS also can be set up to maintain statistics on each task. Maintained are the total time and the minimum and maximum time used by the task. This is set up like this:

```
#use rtos(timer=0,minor_cycle=10ms, statistics)
```

To obtain the statistics, use this code:

```
struct rtos_stats {
   int32 task_total_ticks; // total number of ticks the task has used
   int16 task_min_ticks;   // the minimum number of ticks used in a
                           // single action
   int16 task_max_ticks;   // the maximum number of ticks used in a
                           // single action
   int16 hns_per_tick;     // us = (ticks*hns_per_tic)/10
} stats;

rtos_stats(task_A,&stats);
cout << "Total run time for task_A: " <<
   stats.task_total_ticks/10000*stats.hns_per_tick << "ms" << endl;
```

Summary

- Multitasking is a good way to organize complex programs, breaking them into easier-to-handle tasks.
- Preemptive scheduling allocates a specific time slice to each task and is not practical on most PIC® processors.
- Dispatcher scheduling calls each task in a specific order and is common for PIC® processors.
- A deterministic dispatcher provides a guaranteed execution rate for each task and is considered a real time operating system (RTOS).
- Semaphores are used to control access to a resource.
- Message passing is used to control communication between tasks as an alternative to global data.
- An RTOS can maintain statistics on the CPU time used by each task.

Exercise 23-1

Objective: Learn how to write a program that involves multiple RTOS tasks.
Requires: E3 module, USB cable, PC.

Steps/Technical Procedure	Notes
1. Write a program with the following tasks: Blink red LED at a 1-Hz rate Blink green LED at a 4-Hz rate Display "Button Pressed" when the button is pressed. Use an interrupt handler to capture the button press and a semaphore to transfer the press to the third task.	

Quiz

(1) Of the following operating system types, which most closely matches a Windows OS on a PC?
   (a) Preemptive scheduler
   (b) Dispatcher scheduler
   (c) Deterministic preemptive scheduler
   (d) Deterministic dispatcher schedulers
   (e) PC OSs don't have tasks

(2) Of the following, what is the difference between preemptive and dispatcher type schedulers?
   (a) One allows interrupts and the other does not
   (b) The time a task takes to run is controlled by the task in one and the OS in the other
   (c) The order tasks run in is known in one and not the other
   (d) One supports message passing and it is not possible in the other
   (e) None of the above

(3) Consider a deterministic dispatcher where there are two tasks. One runs at 5 Hz and has a budget of 50 ms. The other runs at 2 Hz and has a budget of 100 ms. What is the smallest amount of CPU time that will be wasted?
   (a) 15%
   (b) 35%
   (c) 55%
   (d) 85%
   (e) None

(4) The following is a list of task frequencies and budgeted time. Which group of tasks cannot be implemented?
   (a) 10 Hz/20 ms, 5 Hz/50 ms, 1 Hz/75 ms
   (b) 20 Hz/10 ms, 10 Hz/25 ms, 5 Hz/5 ms
   (c) 10 Hz/20 ms, 5 Hz/50 ms, 1 Hz/100 ms
   (d) 100 Hz/1 ms, 50 Hz/5 ms, 10 Hz/2 ms
   (e) All are possible

(5) Of the following, what can a semaphore not be used for?
   (a) Count interrupts and control a task execution based on the count
   (b) Manage access to a database so only one task uses it at a time
   (c) As a means for one task to put itself to sleep to be woken by another task
   (d) Task uses it to count down errors and after 10 errors happen the task is suspended
   (e) All of the above are good applications of a semaphore

(6) When using interrupts, special care (usually a disable/enable interrupts) must be taken in accessing multi-byte variables between the ISR and main program. Using a deterministic dispatcher as a scheduler, what special care must be taken to access multi-byte global data used between tasks?
   (a) Enable/disable interrupts
   (b) Semaphores
   (c) Message passing
   (d) Just do not yield until you are done with the data
   (e) Global variables cannot be accessed by tasks

(7) Grouping RTOS functions, which of the following groups does not contain similar functions?
   (a) **`rtos_wait()`** and **`rtos_signal()`**
   (b) **`rtos_msg_read()`** and **`rtos_msg_send()`**
   (c) **`rtos_yield()`** and **`rtos_await()`**
   (d) **`rtos_enable()`** and **`rtos_disable()`**
   (e) **`rtos_overrun()`** and **`rtos_stats()`**

(8) What is the best description for the following code?

```
set_timer_1(0);
rtos_await(get_timer1()>=1000);
```

   (a) The **`rtos_await()`** function is called with 0
   (b) It is a way for a task to safely delay
   (c) The **`rtos_await()`** function is called with 1 only if the code runs very slow
   (d) The code waits until its time slot is up or the timer reaches 1000
   (e) None of the above

(9) What does the following code do?

```
for(i=0, sem=0;i<strlen(line);i++) {
   if(line[i]='(')
      rtos_signal(sem);
   else if (line[i]=')')
      rtos_wait(sem);
}
```

   (a) Stops running if there are mismatched parens on the line
   (b) Pauses after every **`)`** to allow other tasks to run
   (c) Waits on every **`(`** until another task adds data to the line with a **`)`**
   (d) Pauses if there are mismatched parens on the line
   (e) This code will cause an error because the same task cannot wait and signal the same semaphore

(10) Which of the following applications is the best use of a deterministic RTOS?
- (a) A printer processor that uses almost all of its processing time to manage the display, process random incoming data, and perform the printing
- (b) An industrial control panel that must control a pump to be enabled for specific times, manage a user interface, check in with other equipment periodically, and respond to Wi-Fi information requests
- (c) Stop and go light controller that must control lights according to the preset patterns.
- (d) Fire alarm panel that must immediately respond to any sensor change with local alerts and communication with a monitoring station.
- (e) A sump pump controller that will run the motor for 15 seconds when the float switch is activated.

CHAPTER 24

Inline Assembly

Assembly Code with C Code

For those of you who have experience with PIC® assembly language, it may be useful to mix assembly code with a C program that you are writing. Some reasons programmers do this are:

- To directly access special function registers (SFRs). Many programmers think they need to use assembly language to directly access a special function register (SFR). This is of course not true. Here is code to read the PIC® status register:

```
#byte  status = getenv("SFR:STATUS")
....
value = status;
```

- Another reason is there is already a large base of assembly code and now the plan is to switch to C but there is not time to convert all the assembly to C. In this case, the best method is to structure the assembly as a group of subroutines. Assemble that code with your assembler into a relocatable object file. Then in C use the following directive to import the relocatable object file so the functions can be called in C. Finally write your main program in C and call the assembly functions as needed.

```
#import( file=asmfile.o, COFF )
```

  The COFF format is the standard file format for relocatable object files. Relocatable means that nothing in the file has a fixed RAM or ROM address yet. The assembly functions are allocated RAM and ROM locations by the compiler. Symbols (like variables and subroutine names) are exported by the assembly code so they can be accessed in C just like C variables and functions.
- Porting code from another C compiler that does not have built-in functions to access the SFRs because they may have inline assembly to provide access to the hardware. In this case, it is well worth the time to figure out what the assembly is doing and replace it with built-in functions. This way when the project migrates to a new processor, as most projects eventually do, the code is not dependent on specific SFRs.
- The best reason to use assembly in your C program is to optimize a situation where timing is critical or to access special MCU instructions that do not have a direct C equivalent.

Embedded C Programming. http://dx.doi.org/10.1016/B978-0-12-801314-4.00024-7

If you have a really neat trick that you have created in assembly language and have not figured out a way to do it in C, or think it cannot be done in C, then inline assembly may be the answer.

This chapter will cover the details of doing inline assembly. The PIC® assembly language will be covered in a non-comprehensive way just to give the reader a good feel for the kinds of instructions the PIC® devices have.

Inline Assembly Code

The term inline assembly refers to a method of switching from C to assembly and then back, right in your C code. There is no C standard for doing this. Some compilers use a function call like this:

```
asm("MOVLW 5");
```

to insert a line of assembly. Others, like the CCS compiler, use a preprocessor directive to switch to and from assembly. Two preprocessor directives are used:

```
#asm
#endasm
```

The code between the directives is treated as assembly code by the compiler. There is one instruction that is in perhaps all assembly languages. It is the no operation instruction called *NOP*. It does nothing but take up time and space. The keyword "nop" is called a mnemonic. Most instructions have one or more data items after the mnemonic, called operands. The following code inserts three **nop**s in this function:

```
void waste_of_time(void) {
   #asm
      nop
      nop
      nop
   #endasm
}
```

PIC16/PIC18 Simple Move Instructions

The processors in this class have a single working register, W. The assembly instruction set refers to RAM as file registers, or just F. The basic move instructions can only move from a RAM location to W or W to a RAM location. They look like this:

```
movwf  address
movf   address,w
```

The first moves W to RAM and the second RAM to W.

PIC18 only has a special instruction that can move from RAM to RAM. It requires two words in ROM because it needs the extra bits for both RAM locations. It looks like this:

```
movff address,address
```

Accessing C Variables from Assembly

In the inline assembly the absolute RAM address may be specified like this:

```
movf 0x23,w
```

This moves the contents of RAM location 0x23 to the W register. To access a C variable you need only include the name. For example:

```
int8 a,b;
#asm
   movf   a,w
   movwf  b
#endasm
```

The above will move the contents of the variable **a** to the variable **b**. If **a** and **b** were multi-byte variables only the first byte would be moved. Here is an int16 example:

```
int16 a,b;
#asm
   movf   a,w
   movwf  b
   movf   a+1,w
   movwf  b+1
#endasm
```

In both examples, the expression **a** or **a+1** is not the contents of a as would normally be expected. Inside a **#asm** block the variable reference is always an address (like **&a** in C). The **+1** is then adding one to the address not the contents of **a**. Also be aware the addition is not two bytes as would be for **&a** but always a single byte in **#asm**.

It is possible to declare data in a **#asm** like this:

```
#asm
   data 0x12
#endasm
```

Care must be taken so the processor does not execute the data as if it were an instruction (unless that is the intent). Here is data inserted in a safe way:

```
#asm
   goto  lab1
   data  0x12
   lab1: nop
#endasm
```

Labels can be put in assembly just like C. The **goto** is an assembly mnemonic, not the C statement.

PIC16/PIC18 Math Instructions

The primary processor instructions for math operations are listed below. Each allows for the W register and a RAM location to be used as operands. The result can be put into the W register or back to the same RAM location.

```
ADDWF        F+W
ANDWF        F&W
IORWF        F|W
SUBWF        F-W
XORWF        F^W
```

Examples are:

```
ADDWF  A,W      //  W=F+W
ADDWF  A,F      // F=F+W
```

Some instructions set or clear bits in the status register. The two most used bits are the carry (C) and zero (Z) bits. If the result of an add is zero the zero bit is set to 1, otherwise it is set to 0. The carry bit is set if the addition causes a wraparound, such as 255+2, otherwise it is 0. The PIC® data sheet indicates which bits are affected by which instructions.

The following opcode mnemonics use just the RAM location but the result can still be F or W:

```
CLRF     F=0
RLF      F or W = (F<<1)|C   C=F>>7
RRF      F or W = (F>>1)|(C<<7)   C=F&1
COMF     F or W = ~F
DECF     F or W = F-1
INCF     F or W = F-1
SWAPF    F or W = (F<<4) | (F>>4)
```

This opcode just operates on W:

```
   CLRW   W=0
```

PIC16/PIC18 Bit Instructions

The PIC® MCU has some special instructions that operate on a single bit of memory. The *bit clear* (0) and *bit set* (1) instructions are:

```
BCF   address,bit
BSF   address,bit
```

Note that if you have an int1 variable declared in C then you need only the variable name, like this:

```
BSF done_flag
```

The following instructions test a bit and if it is set (1) or clear (0) the next instruction is skipped.

The branch:

```
BTFSC  address.bit      Skip next instruction if bit is clear
BTFSS  address.bit      Skip next instruction if bit is set
```

For example, the following code will set `A` to `0` is bit 3 in the variable `B` is a `0` and to a `1` if it is a 1:

```
CLRF   A
BTFSC   B.3
INCF   A,F
```

PIC16/PIC18 Branch Instructions

The following are the common control instructions:

GOTO address	branch to the ROM address
CALL address	push the next ROM location on the stack and branch
RETLW value	copy the value to W and pop the address off the stack and branch to it. PIC18 has a RETURN instruction that is the same but there is no copy to W
RETFIE	return from interrupt
SLEEP	put the chip to sleep
DECFSZ address	decrement the data at address and if it is zero skip the next instruction
INCFSZ address	like the above except it is an increment; only for PIC18.

The PIC18 has more branch instructions, such as BRA address, which is like a GOTO except it has a smaller range for a short jump. It also has branch instructions based on status register bits, like BZ address that branches only if the **z** bit is set or BNZ address to branch when it is not set.

PIC16/PIC18 Literal Instructions

There are several instructions that use a constant data as the operand instead of an address. These instructions use the W register and the constant only.

ADDLW constant	W=constant+W
ANDLW constant	W=constant & W
IORLW constant	W=constant | W
MOVLW constant	W=constant
SUBLW constant	W=constant−W (not on 12-bit parts)
XORLW constant	W=constant ^ W

Compiler Modifications to the Assembly

Normally in assembly there is a one to one correlation between the assembly and machine code. The compiler, however, does modify the assembly a bit by default. Consider the following instruction:

```
ADDWF   A,F
```

The operand, the variable **A**, is replaced with the address of **A**. The machine code for this instruction on a 14-bit part is:

```
0001111fffffff
```

The **f'**'s are replaced with the address of **A**. Notice there are 7 bits available for the address. If **A** is in location 0x30 then it is OK. However, if **A** is in location 0X93 it will not fit. The PIC® device uses memory banking to solve the problem. Some bits in the status register are used to save the upper bits of addresses that will not fit into the operand field. *Bit 7* of the operand is in status register, *bit 5* on a 14-bit part and the status register is at location 3. The add with a 0X93 operand then looks like this:

```
BSF 3.5
ADDWF 0x13, F
```

When you use variable names in the operand field the compiler adds code to do the extra **BSF**s to access the variable. It will also remember the setting of the status register bits so it knows when to clear them or not set them again.

Some users do not like the compiler changing the assembly, so there is an option to tell the compiler to just process the assembly as it is written. It looks like this:

```
#asm  as-is
   BSF      3.6              // This code adds three to the
   MOVF     0x24,W           // data at 0x124
   ADDLW    3
   MOVWF    0x24
   BCF      3.5
#endasm
```

When using the **as-is** then clear any bank select bits you set when you are done because the compiler assumes they are all reset at the **#endasm**.

SFR Access

In assembly language usually all the SFRs are predefined by name. This is not so in the C compiler. You can always access an SFR by its address, but this is bad form. One option is to define the SFRs used. For example:

```
#byte status=3
```

Then it is possible to do:

```
BSF status.5
```

Better to use the compiler to look up the address, like the following:

```
#byte status=getenv("SFR:STATUS")
```

For a full list of predefined SFRs, from within the IDE, go to VIEW > SPECIAL REGISTERS, and then click on MAKE INCLUDE FILE. It will create an include file with all the definitions.

About the FSR

The RISC instruction set of the PIC® MCU is quite simple. The PIC16/18 parts do not have any fancy addressing modes to access RAM. They do have a feature to allow for indirect addressing. From a C perspective, indirect addressing is like ***x**.

There is an SFR called the *file select register* (FSR). You can load the FSR with an address and then another SFR called the *indirect file* can be used to access data that the FSR points to. The following code will write to **A** indirectly:

```
#asm
   movlw    A                      // W=Address of A
   movwf    getenv("SFR:FSR")      // FSR=W  (address of A)
   movlw    0x12                   // W=0x12
   movwf    getenv("SFR:INDF")     //  *FSR=W   or  A=0x12
#endasm
```

What Not To Do

Modify SFRs with care. The compiler does not look at the assembly to see what you have done, so it may assume a specific setting of the SFRs.

Use absolute RAM addresses with great care. Make sure other functions are not using the RAM locations you pick. It is always best to use C variable names.

Do not make a **GOTO** or **CALL** outside the function unless you plan to abandon your program (like to jump to a bootloader). Doing so will confuse the compiler's method of sharing local variables between functions.

Optimized Assembly

A cyclic redundancy check (CRC) calculation is superior to a checksum; however, the calculation is much more complex. The CRC uses advanced mathematical principles to create a check-sum-like number that will change even when only the order of the data changes. It involves a lot of bit manipulation and that makes it better suited to well-optimized assembly. The following is part of a CRC calculation in C optimized for a PIC®:

```
do {
  for(bit_counter=0x80; bit_counter != 0; bit_counter>>=1)
  {
    done=!bit_test(crc_byte,7);
    crc_byte <<= 1;
    if(*current_data^bit_counter)
      bit_set(crc_byte,0);
    else
      bit_clear(crc_byte,0);
    if(done)
      continue;
    else
      crc_byte ^= pattern;
  }
  current_data++;
} while(--length!=0);
```

Here is the inline assembly carefully optimized for the same CRC algorithm:

```
#asm
 loop1:
  movlw 0x80
  movwf bit_counter
 loop2:
  rlf  crc_byte,f
  rlf  done,f
  movf  current_data,w
  movwf fsr
  movf  bit_counter,w
  xorwf indf,w
  bcf  crc_byte,0
  btfsc zero
  bsf  crc_byte,0
  btfss done,0
  goto  loop2
  movf  pattern,w
  xorwf crc_byte,f
  rrf  bit_counter,f
  btfss zero
  goto  loop2
  incf  current_data,f
  decf  length,f
  btfss zero
  goto  loop1
#endasm
```

On a PIC16 class PIC® the assembly version takes 22 ROM locations and the C version takes 37 ROM locations. As noted, the C code was already optimized to use the special built-in functions for the PIC®. If a published C algorithm was compiled using just standard C operators, the code size will go to 76 ROM locations. Of course, code readability must also be taken into account when using inline assembly. For example, a function only called at start-up and shutdown might not be worth doing in assembly. A function used constantly to transfer data while running, however, might be worth it to improve the throughput.

PIC24 Instructions

The 24-bit PIC® MCU architecture is very different from the other PIC® devices. The PIC24 family has 16 working registers. A few of the registers have a special purpose, so they are not available for general use. The much larger opcode allows for more complex instructions and addressing modes. There are too many details to cover here, but the following shows a few instructions to highlight the differences.

There are many instructions that are similar to the PIC16/18 instructions, previously shown, that only operate on the **W0** register. The following are examples:

```
add   W0,address        // address=W0+address
add   address,W0        // W0=W0+address
```

In addition, for **add** there are the following variations; **Wx** means any W register, **W0-W15**:

```
add  address,Wx
add  Wx,#n,Wx              // Wx=Wx+n     n may be 0-1023
add  Wx,Wa,Wb              // Wb= Wx+Wa
add  Wx,#n,Wa              // Wa=Wx+n     n may be 0-31
```

The **Wa** and **Wb** above can be **W0-W15**, or it can have this form for indirect access:

```
[Wx]               // Data at address Wx
[Wx++]             // Data at address Wx and then increment Wx
[++Wx]             // Increment Wx, then use the data at address Wx
[Wx--]             // Data at address Wx and then decrement Wx
[--Wx]             // Decrement Wx, then use the data at address Wx
```

Normally the instructions operate on 16 bits at a time. Most instructions have a byte mode to only work on 8 bits. To do this a **.b** is address to the opcode. For example:

add.b address,Wx

There are some move instructions that can operate indirectly off a register and offset. These can be used to access data on the stack (only for a move). The stack pointer is always **W15**. For example:

```
mov   [Wx+#n],Wy
mov   Wy,[Wx+#n]
```

There are a few interesting instructions like:

```
inc2  address        // address=address+2
cp    Wa,Wx          // Compare Wa to Wx and set status bits
cp0   address        // Compare data at address to 0 and set status bits
```

A very interesting instruction is the repeat. It repeats the following instruction a certain number of times and then the instruction is executed once more. For example:

```
repeat  #5                  // Executes 6 nops
nop

repeat  #3                  // Adds 2048 to W1
add  W1,#512,W1
```

dsPIC® Instructions

The dsPIC® parts (dsPIC30 and dsPIC33) have special instructions to perform math operations useful for vector arithmetic. All the PIC24 instructions are available on the dsPIC. The dsPIC is considered a digital signal processor (DSP). In general, there is no way to access these capabilities in C directly, so inline assembly is required to use these processor features. Here is an example instruction called the "square and accumulate"; it is a variation on the "multiply and accumulate" instruction and that is where the name comes from.

```
MAC W4*W4, B, [W9+W12], W4, [W10]-=2, W5
```

This instruction does the following:

```
B=B+W4*W4                // There are two DSP working registers A and B
W4=[W9+W12]
W5=[W10]
W10=W10-2
```

First it should be mentioned that although the instruction looks like there are a lot of variables, only certain register numbers can be used in certain spots and in certain combinations. The above does an impressive number of operations (if this is what is needed to be done) and it takes only one instruction time to do it all. The A and B registers are 40 bits and can be set up as either an integer or a fraction from −1 to 1.

Summary

- Assembly functions assembled with an assembler into relocatable object files can be called from C.
- Assembly code can be inserted into C functions.
- Inline assembly can access C variables by name.
- The compiler by default adds instructions to facilitate bank switching for RAM access.
- PIC24 instructions are very different to and much more powerful than PIC16/18 instructions at the assembly level.
- dsPIC-specific instructions must be coded in assembly.

Exercise 24-1

Objective: Gain experience writing inline assembly inside a C function.
Requires: E3 module, USB cable, PC.

Steps/ Technical Procedure	Notes

1. Complete shell function to count the number of 1 bits in a byte. Write a program to test the operation of the function. The function shell is:
`int8 number_of_one_bits(int8 data){`
`  int8 count;`
`  #asm`
`   // to do`
`  return count;`
`}`
HINTS:
• In addition to transmit and receive it is a good idea to connect the grounds between the two boards.	

Quiz

(1) When might assembly language be a good idea?
  (a) To access SFRs
  (b) To help in portability
  (c) To code a time-critical algorithm
  (d) To directly write to I/O ports
  (e) To improve readability

(2) Which of the following statements is false?
  (a) C can call assembly functions
  (b) Assembly functions can call C functions
  (c) Assembly code can be inside C functions
  (d) Assembly code can access C variables
  (e) Assembly code can have C labels

(3) On a PIC16/18 class part, how many instructions inside the **#asm** are required to perform the byte operation **x=x&5**?
  (a) 1
  (b) 2
  (c) 3
  (d) 4
  (e) 5

(4) On a PIC16/18 class part, how many instructions inside the **`#asm`** are required to perform the byte operation **`x=x|4`**?
   (a) 1
   (b) 2
   (c) 3
   (d) 4
   (e) 5

(5) On a PIC device, when the destination appears in the operand list, where is it in the list?
   (a) The first operand
   (b) The last operand
   (c) It is always part of the opcode
   (d) It depends on the opcode
   (e) It is never in the operand list

(6) Of the following, which will not work to access the status register on a 14-bit part?
   (a) **`clrf status`**
   (b) **`clrf getenv("SFR:STATUS")`**
   (c) **`clrf 3`**
   (d) **`clrf 0b00000011`**
   (e) **`clrf 1+2`**

(7) In the PIC16/18 mnemonics, the letters L, W, and F refer to what?
   (a) Label, W register, file register
   (b) Constant, working register, fraction
   (c) Last result, W register, first result
   (d) Constant, working register, RAM location
   (e) Literal, W register, fraction

(8) There is a PIC® instruction not mentioned in this chapter. It is **`MOVF address, F`**. Based on what you already know of PIC® mnemonics, what does this instruction most likely do?
   (a) Moves W into a RAM location
   (b) Moves a RAM location into an address pointed to by the FSR
   (c) Sets the status register based on the value at the RAM address
   (d) Moves a RAM address into W
   (e) There could not be an instruction like this

(9) For a processor with a 7-bit field in the operand, 256 RAM locations, and 32 SFRs, how many bits are needed for blank selecting?
  (a) 1
  (b) 2
  (c) 3
  (d) 4
  (e) None

(10) The following assembly was written to implement **`if(flag) x=0;`** on a 14-bit processor. Flag is in the third bit of 0x0C and **`x`** is at 0xFF. What is the problem with this code?
  (a) The address of flag is wrong
  (b) There is a missing operand on the clrf
  (c) The clrf will always be executed
  (d) The wrong bit is being cleared at the end
  (e) There is nothing wrong with the code

CHAPTER 25

Debugging

Overview

So far in this book we have used the E3 board tethered to PC with a preloaded bootloader to load and run all the programs. This is an expedient way to learn the language and processor; however, in the real world you are not likely to put the E3 board in a washing machine. This chapter deals with debugging programs. Many of the techniques in this chapter cannot be used without debugging hardware and are therefore more suited to real products as opposed to the educational E3 board.

Simulators are used to expedite testing debugging of computer programs. Although simulators are available for a PIC®, including a free one from Microchip, they are not very popular for embedded systems. A simulator is software running on a PC that can load a hex file and will execute the program the way the processor would do it. The advantage is you can stop the execution, step one instruction at a time, and view memory to help test and debug. The problem with simulators for embedded use is you don't have a real interface to your hardware. For example, when writing code for a microwave oven, your program interfaces to an LED display, keypad, various switches/sensors, and more. To set the simulator up to properly simulate your interfaces can be a lot of work and in the end there is nothing like testing on the real hardware.

Emulators are considered the gold standard in debugging. An emulator is a hardware unit that replaces the processor in your product. It connects to a PC and allows the same kind of capability as the simulator. You can load a hex file, stop the program, look at memory, and much more. With an emulator the code can run in real time and you get all the normal hardware interfacing. Emulators are expensive and you need one for your specific processor.

Modern PIC® processors have debug capability built into the chip. An inexpensive in-circuit debugger (ICD) unit can be used to get many, but not all, of the debugging capabilities of an emulator. This has become the most popular method of debugging PIC® processors.

Most products, even in production, have a programming connector on the board so the chip can be reprogrammed at any time by connecting up a device programmer. This is called in-circuit serial programming (ICSP). The value in this can be great if a program needs to be

Embedded C Programming. http://dx.doi.org/10.1016/B978-0-12-801314-4.00025-9

changed because a bug was discovered. Boards and units not yet shipped can be updated easily and for more expensive equipment field service technicians can update units in the field. This can also be used for recalled devices to prevent just throwing out the processor board. The same connector can be used for ICD debugging. This chapter will start covering that kind of debugging and then move on to alternate techniques.

ICSP

In-circuit serial programming requires three pins on the PIC®. In addition, the Vdd and ground are connected to the device programmer. The CCS prototype boards and many Microchip boards also connect up a sixth pin. This can be used in debugging but is optional.

The limitation of ICSP is that two of the processor I/O pins are used for the programming connector. The connections required to the PIC® are detailed in the following paragraphs.

Vpp (aka $\overline{MCLR}$/Vpp): the Vpp pin is used on the PIC® device to put the chip into programming mode. On most chips this is done by applying a voltage over 5 V (usually from 9 V to 13 V depending on the chip). On most parts this pin is shared with $\overline{\text{MCLR}}$. It is normal to have a pull-up resistor on $\overline{\text{MCLR}}$ to allow the program to run. For programming, if the resistor is too small, it will slow the rise time of the high voltage and that will prevent the chip from entering programming mode. CCS recommends a 47K pull-up on this pin. Microchip data sheets and application notes vary from 1K to 47K on this resistor. We have seen where some parts are fussy about this rise time and there are even differences in the part lots. The 47K seems to work reliably on all parts.

Sometimes you will see a diode and/or capacitor on this pin. The diodes are OK (between the pin and resistor); however, the capacitor again can affect rise time and it may interfere with programming.

Some newer parts do not use a high voltage on the pin, they just use Vdd and some special pulsing. The 47K still works well on those parts.

PGC/PGD: these pins are the programming clock and data. In fact programming is done using an SPI bus. On many parts these pins are B6 and B7. Check the data sheet for the pins on your part. The names do vary a bit. For example, ICSPC and ICSPD.

Do not use series resistors on this pin as they may interfere with the bidirectional communication. Less than 30 ohms may be okay. Small caps 47pf and under are also okay on these pins if noise reduction is needed. Usually nothing is put on these pins.

Because these are also I/O pins, some designers will connect these pins to other hardware on the board. This can cause trouble. We want as light a load as possible on these pins. Add jumpers if needed to select programming mode or run mode.

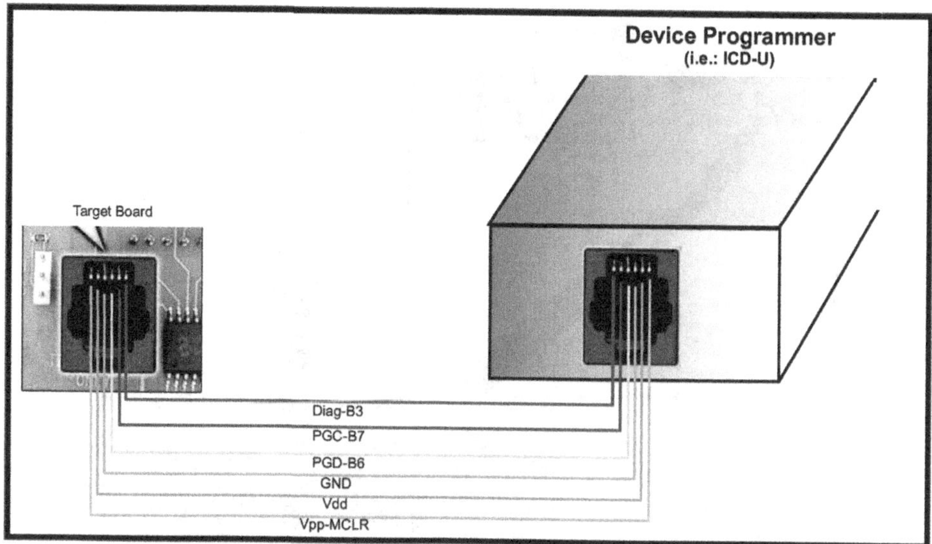

Figure 25.1: ICSP target to programmer connection diagram.

ICSP Jacks

The target jack on most programmers is a modular phone jack with six wires. Sometimes the same jack is used on the target board. When this is done the cable should mirror the pin numbers. Pin 1 goes to pin 6. Figure 25.1 shows the standard pinouts.

Sometimes the programmer powers the board and sometimes the board supplies Vdd to the programmer. Make sure you know what configuration you are using. Either way, Vss and Vdd must be connected from target to programmer.

Another popular connector on the target board is a 1x6 (or 1x5) pin header. Adapter cables are available that go from modular jack to five- or six-pin header. These take less board space and are less expensive than the modular jacks.

Connectors on every board produced can get expensive. This is especially true if another reprogramming is ever needed. An alternative is to use a connection method that does not require extra hardware. A card edge connector can be used for this. On the PCB side this is just some extra traces and routing.

Another option is to use what is called a tag-connect connector. On the programmer side it has a modular jack to the programmer, and on the target side several spring-loaded pogo-stick-like connectors with metal guides and plastic hooks. On the PCB you simply need a pad and hole pattern to match up with the pogo sticks and guides. There is no additional cost to

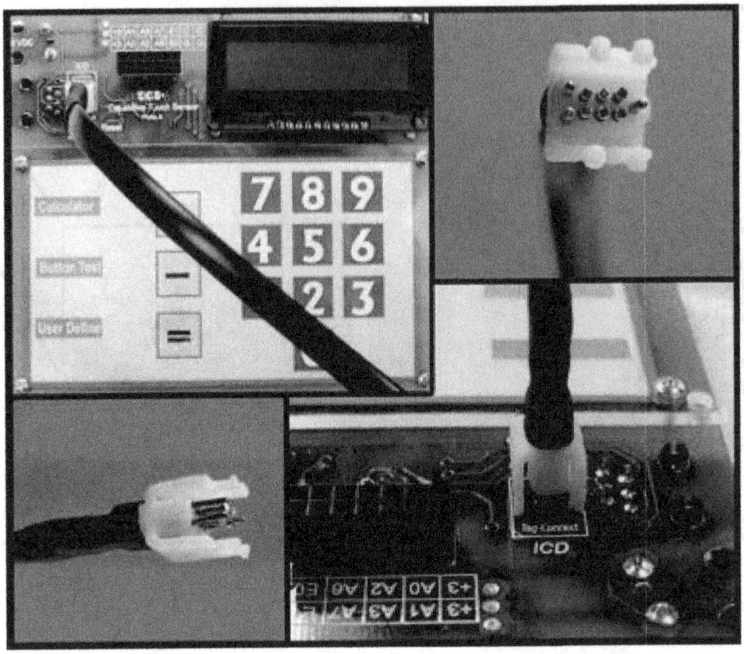

Figure 25.2: Tag-connect socketless ICSP photo.

the PCB (product) and it is a small connector so not much board space is used. Figure 25.2 shows what this looks like.

Breakpoints

The debugging hardware will have a way to freeze program execution when a specific address is executed. With a PIC® using an ICD, the instruction at the address of the breakpoint is executed so the program counter points to the next instruction when it stops. The processor does not actually stop. It jumps to a small debug executive program loaded into the chip with your program. That code communicates with the external ICD unit through a serial peripheral interface (SPI)-like interface. The ICD can request reads and writes to the chip memory to help the PC debugger to do its job. The ICD can also command the debug executive to return control back to the program.

One complication you will need to be aware of is what happens to the peripherals when program execution stops. Many PIC® processors will let you configure the debugger to either keep them running or to freeze them. For example, if a timer is running when the breakpoint you will be able to decide if the timer should stop incrementing while the application program is stopped in the debugger. Most frequently you will want the peripherals to stop.

In the IDE, to start debugging make sure the ICD unit is connected to the PC and click on DEBUG>ENABLE. At this point you can manually start and stop program execution from the debug control panel.

To set breakpoints you can double-click in the left gutter on the source code editor where you want the breakpoint set. Another double-click will remove the break. The debugger break panel will show all the breakpoints and you can manage the breakpoints from there as well. You can also set a breakpoint on an assembly instruction by opening the C/ASM (.LST file) view.

There is another thing you need to be aware of, because the PIC® is executing the line it breaks on. The following shows the C/ASM for a program to help us see what happens when breaking in the C source.

```
....................  pos2='1';
0076:  MOVLW  31
0077:  MOVWF  35
....................  counter++;
0078:  INCF   36,F
0079:  BTFSC  03.2
007A:  INCF   37,F
....................  output_b(0);
007B:  BSF    03.5
007C:  CLRF   06
007D:  BCF    03.5
007E:  CLRF   06
```

When a breakpoint is set on **counter++;** in C then the debugger sets the breakpoint at address 0078. When execution stops then the next address to execute is 0079. In C the green execution pointer will be at **counter++;** as if that were the next line to execute. In reality the first assembly line of that line is already executed. Counter is an int16 (2 bytes) and the low byte was already incremented when the break was hit. If you look at the counter value you will see it was incremented even though from the debugger view it seems like the increment line was not yet executed. This is an annoyance we need to put up with using the inexpensive ICD method of debugging. Notice if we set a break on either of the other lines shown here you would not notice or care that the first assembly instruction was executed.

It can help to switch to the C/ASM view to see where the green arrow is in that window. There you will see the green arrow on line 0079 after breaking at **counter++;**. In that view you also have the option to set a breakpoint on 0077 to get it to stop just before the increment.

The number of concurrent breakpoints that can be set will vary depending on the chip.

Some chips also offer data breakpoints. A data breakpoint allows you to stop execution when there is a read and/or write to a specific RAM location. This is a powerful debugging feature on the chips that have it. To set a data breakpoint use the break panel in the debugger.

Viewing Memory

While program execution is stopped the debugger can show the contents of RAM. There is no way to view memory while the program is running. The easiest way to check the value of a variable is to hold the mouse cursor over a variable in the source code. The debugger pops up the value in a small window. This is called mouse-over. See Figures 25.3 and 25.4.

The debugger also has a watch panel where you can enter any number of C expressions. Those expressions (usually just a variable name) are evaluated and the values shown in the watch panel. You can change the radix each watch item is shown in (for example hex or decimal).

In addition to the watch panel, the debugger has RAM panel that shows the entire contents of the RAM. The special function registers (SFRs) are included in the RAM view and they are shown by themselves by name in the SFR panel. Finally, you can view the SFRs sorted by peripheral and with annotations as to what the registers mean using the peripherals tab. See Figure 25.5.

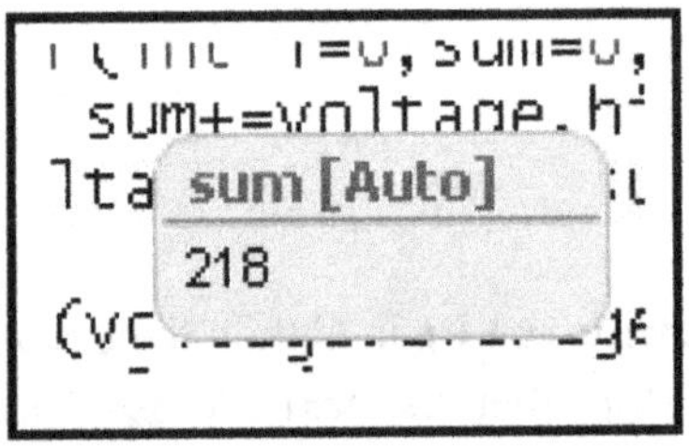

Figure 25.3: Simple mouse-over pop-up.

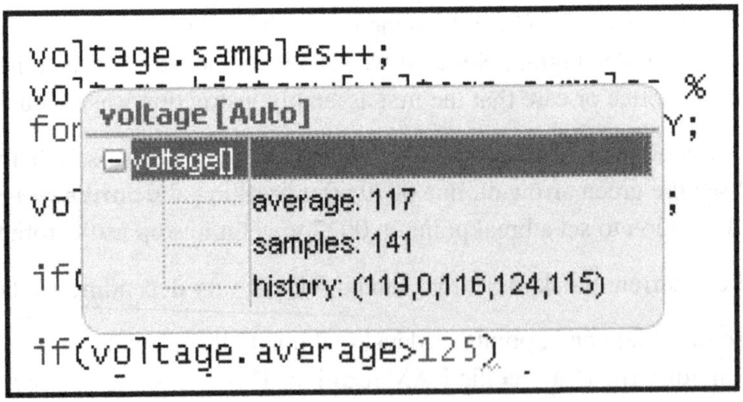

Figure 25.4: Structure mouse-over pop-up.

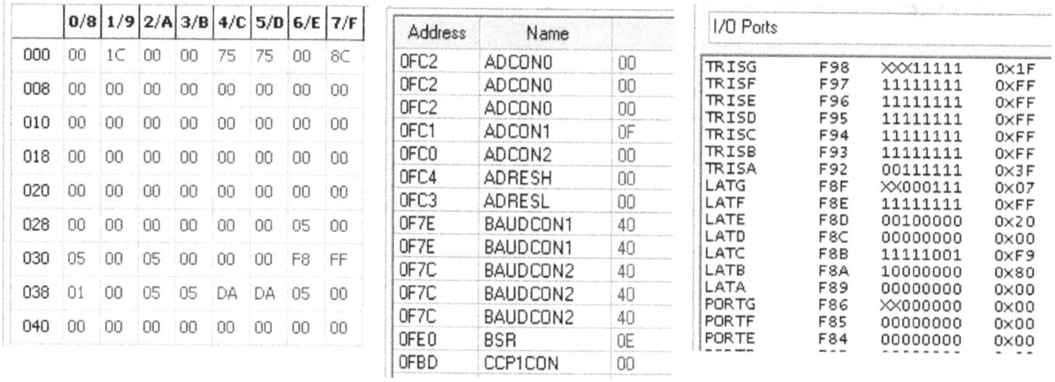

	0/8	1/9	2/A	3/B	4/C	5/D	6/E	7/F
000	00	1C	00	00	75	75	00	8C
008	00	00	00	00	00	00	00	00
010	00	00	00	00	00	00	00	00
018	00	00	00	00	00	00	00	00
020	00	00	00	00	00	00	00	00
028	00	00	00	00	00	00	05	00
030	05	00	05	00	00	00	F8	FF
038	01	00	05	05	DA	DA	05	00
040	00	00	00	00	00	00	00	00

Debugger RAM view

Address	Name	
0FC2	ADCON0	00
0FC2	ADCON0	00
0FC2	ADCON0	00
0FC1	ADCON1	0F
0FC0	ADCON2	00
0FC4	ADRESH	00
0FC3	ADRESL	00
0F7E	BAUDCON1	40
0F7E	BAUDCON1	40
0F7C	BAUDCON2	40
0F7C	BAUDCON2	40
0F7C	BAUDCON2	40
0FE0	BSR	0E
0FBD	CCP1CON	00

Debugger SFR view

I/O Ports

TRISG	F98	XXX11111	0x1F
TRISF	F97	11111111	0xFF
TRISE	F96	11111111	0xFF
TRISD	F95	11111111	0xFF
TRISC	F94	11111111	0xFF
TRISB	F93	11111111	0xFF
TRISA	F92	00111111	0x3F
LATG	F8F	XX000111	0x07
LATF	F8E	11111111	0xFF
LATE	F8D	00100000	0x20
LATD	F8C	00000000	0x00
LATC	F8B	11111001	0xF9
LATB	F8A	10000000	0x80
LATA	F89	00000000	0x00
PORTG	F86	XX000000	0x00
PORTF	F85	00000000	0x00
PORTE	F84	00000000	0x00

Debugger peripheral view

Figure 25.5: Debugger RAM, SFR, and peripheral views.

Stepping

Once program execution has stopped, in addition to just starting execution again there are STEP and STEP-OVER buttons that can be used for a limited execution. Step will execute a single line of code and then stop again. If the next line is a function call then the execution will stop at the start of the function called. Be aware the green arrow points to the next line to be executed.

Step-over works like step except when the line to be executed is a call to a function. In that case the entire function is executed and execution stops at the line after the call.

Be aware the steps work differently when the open editor window is the C/ASM .LST file. In this case each step is a single assembly instruction. When a C source file is the active editor window then a C line is executed.

With breakpoints we need to be careful because the first assembly instruction is executed on a break. When stepping we don't have this problem. That is because the debugger analyzes the assembly code for the next instruction and is able to back up the break address so only the current C line is executed. This isn't possible to do with breakpoints due to jumps in the code.

When doing a lot of single stepping you will notice a pause between each step. Using the ICD debugger, the contents of RAM must be serially transferred from the PIC® to the PC each time the debugger stops. If this gets too long, change the debugger view so that only as few RAM locations as are needed are visible. The debugger will only request data it needs to show on the screen.

Power Debugging

The debugger in the IDE has features that can be exploited to make the debug process more efficient. One concept to get used to is the project notepad. The IDE can keep a file for each project where you can record notes. Many of the debugger panels can be made to use this file.

For example, you can set up the debugger to record certain variables, RAM locations, and a lot more each time the debugger stops the program. That includes user halt, breakpoints, and single stepping.

The debugger can also be set up so only this data is logged when a breakpoint is hit and the the program execution continues. This can be use to dump certain variables each time a specific line in C is executed.

Another use is to highlight an area in the RAM map and right-click to copy the data to the notes file.

Monitor

The ability to log data each time a certain line is executed as described above is great; however, sometimes you need a more powerful data interface to the program. The ICD can be used to transfer user data back and forth to the running program. The way this works in C is to define a special serial stream, like this:

```
#use RS232( debugger,  stream=dbg )
```

In the program you can then have lines like this:

```
dbg << "Gallons=" << precision(2) << total_gallons << endl;
```

You can even do input like this:

```
dbg << "Enter delay time to use: ";
dbg  >> delay_time;
```

The data appears in the monitor panel of the debugger. It is in this panel that you also can type data to be sent to the PIC®.

There are some limitations, however. First the ICD unit needs an additional wire connected to the PIC®. The normal modular jack used for many ICD connectors already has the sixth pin available and many hardware designs already connect it to the PIC®. If you don't use the default pin (B3 on PIC16 and B5 on PIC18) then you need to change the **#use** directive like this:

```
#use RS232( debugger,  rcv=PIN_B2, xmit=PIN_B2, stream=dbg )
```

The extra wire is used to transfer data in both directions to the PIC®. Because the same wire is used in both directions, be aware there is no hardware buffer for the data so the program must be waiting for incoming data when you type at the PC.

Data Streaming

The monitor feature covered above is a powerful way to interact with your running program. It only works, however, when the debugger is running. The ICD can be used outside of a debugging environment to simply transfer serial data between the PIC® and the PC.

The CCS toolset calls this data streaming. In this configuration only the normal programming pins on the PIC® need to be connected. The extra pin used for the monitor feature is not required.

Imagine you have a PIC® processing data for TV remote receiver. In your code, each time you get a command you might output the raw data received out of a data streaming port. All you need to do is connect up and ICD and you will get a live dump of the data at the PC to help diagnose problems.

This interface can even be left in the production code in case a field service person needs to diagnose a problem at a customer's site. Because the data is bidirectional like the monitor data, you can require a password to enable the data dump.

It can also be used to set calibration data, serial numbers, or date of manufacture in a production environment.

The code looks a lot like it did for monitor. Here is some C:

```
#use RS232( icd,  stream=dbg )
...
dbg << "Enter calibration value: ";
dbg  >> cal_data;
```

By default the pins used are the ICD PGC and PGD pins; however, data streaming will work on any pins. If the pins you use are UART pins then you also get the benefits of a UART in your communication; for example, interrupts.

Real-Time Issues

In the world of computing, the debugging tools like breakpoints and single stepping are basic and commonly used debugger features. For some embedded programs, however, they cannot be used.

Here are some simple examples to demonstrate the issues:

- Controller for window blinds. If you hit a breakpoint while closing, the motor will just keep running and there is no program running to stop it.
- TV remote receiver. Hit a breakpoint and it will stop the code but that will not stop or slow down the transmitter. You can examine the data from the first break but there will be no way to continue.
- HVAC motor speed control. The program may need to respond to many interrupts every second just to keep the motor operating correctly. One break and the motor breaks.

You will find many more examples similar to the above. Virtually every program that is in active communication with other devices will have this problem. It is not uncommon for multi-processor systems to use one processor to shut down the whole system if it appears another processor has stopped responding.

This is not to say it is impossible to set a breakpoint. You can modify the devices the program talks to and use hardware simulators instead of real hardware for dangerous interfaces. There will be a moderate amount of work for some capability, but you should be aware that many of the problems you need to debug will only happen with real hardware and even then probably infrequently.

In addition to data streaming the following few sections have some other techniques for debugging that can be helpful in situations where you can't use breakpoints.

Use of a Scope

A simple oscilloscope can be a huge help to gain some insight into what is going on in your code. You will need to find one or more unused pins that can be connected to a scope. You may be able to use PGD and PGC if you aren't using data streaming or the debugger.

Here is a simple example:

```
#int_ext
void  isr(void) {
   output_high(PIN_B6);
   // normal processing here
   output_low(PIN_B6);
}
```

This simple modification of two lines in the interrupt service routine (ISR) will cause pin B6 to be high while the interrupt function is processing data. This will give you on the scope screen the exact time the interrupt takes to process data. In addition you could use another scope channel to monitor an external stimulus like a serial data signal in, PWM input, or simple push button and compare that data to when the interrupt is processed to analyze latency and other characteristics.

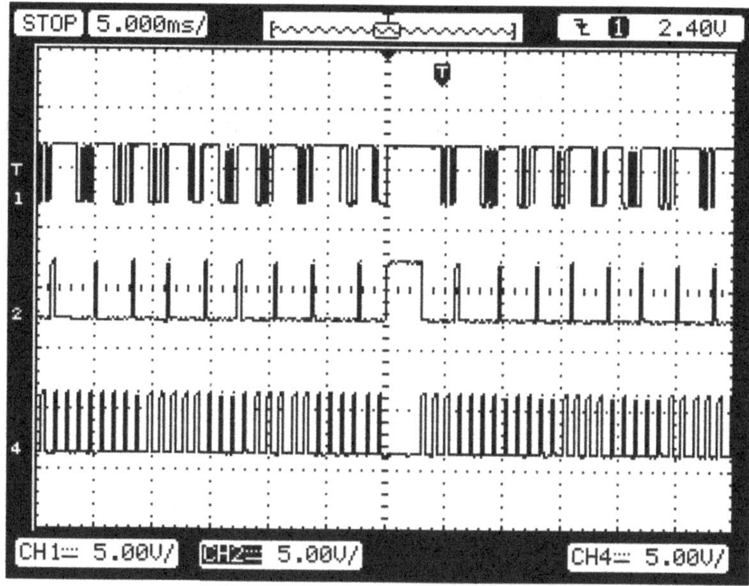

Figure 25.6: Using a scope as a debugging tool.

Multiple pins can be used to see the relationship between when various areas of the code are executed. A nice characteristic of this kind of debugging is the program timing is barely changed by the debug code. Adding printf's to your code can make enough of a timing change to either break your code or fix broken code. Review the example trace shown in Figure 25.6.

The top trace is asynchronous serial data coming into the processor. The middle trace is high while in the RDA interrupt function reading the byte. The bottom trace is a timer interrupt that fires off every millisecond.

Notice at the end of each byte the RDA fires off quickly except for 1 byte on the screen that takes a long time to process. This is the end of packet and more processing is required to verify the checksum and queue up the data. The problem is the timer interrupts (bottom trace) are lost during this processing.

This scope trace makes it very easy to see that timer interrupts are being lost.

Here are some examples of how this technique can be used:

- Some interrupts are being serviced too late and it is suspected the problem is excessive interrupt latency. Use different diagnostic pins to track each interrupt to figure out if any of the interrupts are active when the problem interrupt should fire. If that fails to find the problem, use the same technique anywhere in the code where interrupts are disabled.
- Hardware sends pulses to the processor and an interrupt should fire for each pulse. The count in program does not match what is being sent. Put the ISR activity on one channel

and the pulse input on the other channel. Look to see if a pulse is missed and then examine the pulse signal with the scope to see if there is something wrong with it.

- Occasionally the program detects an inconsistency in the hardware signals that should never happen. Track each signal on a scope and use a diagnostic pin as the scope trigger. In code set the pin high when the error condition is detected. The scope data should then tell what is going on.

If not enough spare pins are available you may need to run several tests to get all the data. Another technique is to do something like this:

```
#int_ext
void  isr(void) {
    output_high(PIN_B6);
    delay_us(25)
    output_low(PIN_B6);
    delay_us(25);
    output_high(PIN_B6);
    // normal processing here
    output_low(PIN_B6);
}
```

The 25-us high then low could be a different time in every place B6 is used and the scope can be used to identify what function it was based on the time. This of course adds 50 us to the ISR time. Smaller times can be used if the scope can resolve the times with the needed time range.

Diagnostic Interface

Many products will have some kind of diagnostic port designed into the product. The above data steaming method may be used for these products. When a more complex interface is needed there will typically be a special PC program to talk with the product. Some kind of key may be sent to enable the interface and to set the level of diagnostics. For example, the interface might allow for certain sets of diagnostic data to be enabled. It may have the ability to read and maybe even write to RAM. Service software may be used to put the unit into test modes or to just monitor the system operation.

Record/Playback

Sometimes you will have a simple main program that reads data and then processes data at some rate. For example:

```
while(TRUE) {
   set_timer0(0);
   gather_data();
   process_data();
   while( get_timer1() < 50000 ) ;  // wait for 100ms mark
}
```

This type of program can be easily modified to record and play back the data, making it easy to investigate problems that are intermittent. The data could be saved in a serial EE or sent to a PC for storage. The program could look like this:

```
while(TRUE) {
   set_timer0(0);
   if( mode==MD_PLAYBACK)
      read_next_data();
   else
      gather_data();
   if(mode==MD_RECORD)
      save_data();
   process_data();
   while( get_timer1() < 50000 ) ;  // wait for 100us mark
}
```

There are many ways to structure a program and its data. It will help in your program design to consider up front what you may need to investigate problems and verify the design.

Profile Tool

The CCS C compiler has an interface to a profiling tool that can be used to efficiently send program execution information out real time to a PC. In your C code you need something like this:

```
#use profile()
#profile  profileout
...
profileout("Got to point A");
```

When the program is run and the above line is executed, the message "Got to point A" appears on the PC screen. The interface used is the ICD, just like data streaming. The advantage of this method over data streaming and a printf is the full string is not sent through the ICD. Only a code is sent, and the full text is part of the debug file used by the IDE. At the PC the messages are timestamped.

Variable values can also be sent with or without a message:

```
profileout(total_amount);
profileout("Grand total", total_amount);
```

Variables are sent in binary with a tag indicating the type, so you don't need the formatting indicators you would have in a **printf**.

Profiling Code

The real power of the profile tool is the ability to have the compiler automatically insert tags to help trace the flow of your program. For example:

```
#profile  functions
// various functions here
#profile off
```

This will cause the compiler to insert a tag at the start and end of every function between the two profile directives. At the PC, while the program is running we get live data each time one of the tagged functions starts or stops. One view will show you the sequence of function calls (see Figure 25.7). Another view shows the count each function is called (see Figure 25.8) and the minimum, maximum, and average execution time.

In addition to the start and stop of each function you can have the function parameters sent on each call. In this case the parameters are shown on the PC call sequence display.

```
RESET, Cause: Normal power up
  init_hardware( 0 )
    reset_filters()
    built_in_test()
  gather_inputs()
  control_logic()
    safety_checks( 5 )
    calculate_area( 112 , 34 )
  set_outputs()
    motor_control()
    led_control()
  update_display()
    lcd_putc( '\0x1E' )
```

Figure 25.7: Profile tool sequencing view.

	Count	Min	Ave	Max
init_hardware	1	215ms	215ms	215ms
reset_filters()	1	2us	2us	2us
built_in_test()	1	137ms	137ms	137ms
gather_inputs()	71	28us	28us	28us
control_logic()	71	233us	233us	250us
safety_checks	71	25us	25us	25us
calculate_area	71	44us	44us	60us
set_outputs()	71	21us	21us	21us
motor_control()	71	17us	17us	18us
led_control()	71	1us	2us	2us
update_display()	71	96us	96us	112us

Figure 25.8: Profile tool function time view.

Another feature that will require more data to be sent with each tag is to have the compiler add the current value of a PIC® timer with each tag sent out. This allows for much more accurate time values shown at the PC. Change the **#use** directive like this:

```
#use profile(TIMER1)
```

Tracing and timing functions can tell you a great deal about how your program is running. When rigorously testing code there is a concept called full path testing. The idea is to make sure your tests cause every line of code in the program to be executed. The profile tool has a feature to make verifying full path testing easier. The following option causes the compiler to add a tag at every possible branch in the code. At the PC, the branches are identified by line number.

```
#profile  functions, paths
```

Design Verification

A test procedure will typically step through all of the requirements for the project and fully test each requirement. Usually it is not possible to test all combinations of all external inputs to fully test a program under all circumstances. A test procedure will have selected tests to cover the most common, some randomly selected, and some special tests. The special tests use what the programmer knows about the internal weaknesses of the program. For example, testing 255 and 0 for a value stored as a byte.

An excellent test procedure will also verify elements of the design that may not trace back to specific requirements or be evident under normal execution. An important part of design verification on embedded systems involves timing. Consider the following example:

- An industrial controller has a laser-based sensor that sends a pulse 10 times a second as long as the operator's finger is not in the way. Once a second the firmware will send an ultrasonic pulse and then reset a timer for a CCP capture. An ultrasonic receiver will trigger the CCP, capturing the time it took for the pulse to bounce off the material in the machine and get back to the receiver. The CCP time represents the distance, and this is shown on an LED panel. Testing seems to show everything works well. After thousands of units have been shipped it becomes clear some units will occasionally show the distance as off by an inch.
- The problem is when the ultrasonic pulse is sent, if the laser interrupt comes in before the timer is zeroed then the time in the safety ISR (around 150 us) is lost and the timer value will be too low when the CCP triggers. Not all units seem to show the problem because once powered up the PIC® timing for the ultrasonic and the laser unit's timing are kind of synchronized. Electronic component values determine exactly how fast each unit powers up and how accurate the clock is. To find this kind of problem you need to have a very aggressive test plan or do a very thorough design review with seasoned engineers.

With a program structure that has a main loop as shown above, temporary code could be inserted to get an idea about the percentage of CPU time that is used, like this:

```
max=0;
count=0;
while(TRUE) {
    set_timer0(0);
    gather_data();
    process_data();
    if(get_timer1()>max)
        max=get_timer1();
    if(++count==10) {
        cout << "CPU used=" << max/250 << "%" << endl;
        max=0;
        count=0;
    }
    while( get_timer1() < 25000 ) ;   // wait for 100ms mark
}
```

Finding the execution time of your interrupts can be very useful in assessing your program's performance. Another temporary technique that relies on a free-running timer looks like this:

```
#int_ext
void ext_isr(void) {
   int16 time;
   time=get_timer1();
   // normal ISR code here
   time=get_timer1()-time;
   ext_isr_time += time;
   ext_isr_count++;
   if(time>ext_isr_max)
      ext_isr_max=time;
}
```

You can stress-test interrupts that might break with a high latency by introducing an artificial latency like this:

```
#int_ext
void ext_isr(void) {
   delay_us(50);
   // normal ISR code here
}
```

Increase the delay time until the code stops working. You then know the maximum latency the code can tolerate. Compare this to processing times found in the other ISRs. It is best to do this kind of analysis while the program is being designed; however, frequently hardware tolerances that may not be known need to be taken into account and this requires testing.

Many of the situations we design our code to deal with are difficult to make happen on demand. To properly test this code you may need force certain conditions to see that the code handles it correctly. For example, you may want to intentionally corrupt a program memory location to test your check-summing or CRC code. Another example would be to force a certain timing situation to ensure the code can handle a worst case situation.

Do not forget in your validation testing to test for abnormal and unexpected inputs. For example, on an I²C interface, grounding the SCL line will cause many programs to hang. Sometimes the watchdog timer will go off and reset the chip and that will be as good a response as any. Other programs will need to put the system into a safe state and/or save data in EE. Opening the SDA line on an I²C can cause some unexpected data on reads and that is another good test of your code's logic.

Summary

- Debugger tools for the PIC® include simulators, emulators, and ICD-type debuggers. The ICD is the most popular and cost effective.

- An ICSP connector can be used for chip programming and debugging, and may be used as a communications port.
- The core capabilities of a debugger are breakpoints, single stepping, and viewing RAM.
- Advanced debug capabilities build on the core capabilities to provide functions such as logging and C expression evaluation.
- The monitor functionality in the CCS C compiler uses the ICD in the debugger mode to transfer **`printf`** and **`getc`** data to/from the PC debugger screen.
- The data streaming functionality in the CCS C compiler uses the ICD in the normal run mode (not debugging) to transfer **`printf`** and **`getc`** data to/from the PC debugger screen.
- Special considerations must be taken with real-time programs because of the complexities of timing and hardware interfacing.
- Use of an oscilloscope or logic analyzer as well as using a spare serial port to dump data helps greatly to debug embedded programs.
- A profiler tool allows for good information about function timing, sequencing, and what code gets executed.
- Design verification for embedded systems often requires modifying the code to obtain timing data and to force conditions that happen infrequently.

Exercise 25-1

Objective: The following program is designed to ask the user for the size of each room in a house. It will then calculate and output the area in cubic feet. This program has one or more flaws that cause it to output the wrong number for the given test case. The object is to debug this program.
Required: E3 module, USB cable, PC.

Steps/Technical Procedure	Notes
1. Test data: 7 0 9 6 10 0 12 0 10 0 11 6 8 8 12 2 41 6 3 0 5 0 0 0	

2. Write Program (also in the examples directory as ex_ch25.c):
`#include <ios.h>`
`typedef int16 inches;`
`typedef int16 cubic_feet;`
`#define MAX_ROOMS 10`
 	
`typedef`
`   struct {`
`      inches height;`	

Steps/Technical Procedure	Notes
(code listing below)	

```
    int8  room_count;
    struct {
       inches width;
       inches length;
       } rooms[MAX_ROOMS];
    } house_type;
int16 get_dimension(rom char*for_what) {
   int feet_in, inches_in;
   cout << "Enter room " << for_what << " in feet,
space, inches:";
   cin >> feet_in >> inches_in;
   return feet_in*12+inches_in;
}

void input_data(house_type*house) {
   house->height=get_dimention("height");
   cout << endl << "Enter data for each room, 0 0 when
one." << endl;
   house->room_count=0;
   do {
      cout << endl;
      house->rooms[house->room_count].width=get_
dimention("width");
      if(house->rooms[house->room_count].width==0)
          break;
      house->rooms[house->room_count].length=get_
dimention("length");
      house->room_count++;
  } while (house->room_count<MAX_ROOMS);
}
cubic_feet calculate_volume(house_type*house) {
   int32 cubic_inches=0;
   for(int i=0; i<house->room_count; i++)
      cubic_inches+=(int32)house->height*
       (house->rooms[i].width*house->rooms[i].length);
    return cubic_inches/(12*12*12);
}

void main(void) {
  house_type house;
  do {
     input_data(&house);
     cout << "Volume of air in house=" << calculate_
volume(&house)
         << " cubic feet" << endl << endl;
  } while(TRUE);
}
```

(continued)

Steps/Technical Procedure	Notes
HINTS: • Add **`printf`** statements to find the problem. • The correct answer is: 5842.	

Quiz

(1) When setting a breakpoint on a **GOTO** assembly instruction, where does the processor actually stop?
   (a) The **GOTO** instruction
   (b) The instruction before the **GOTO** in memory
   (c) The instruction executed just before the **GOTO**
   (d) The instruction after the **GOTO** in memory
   (e) At the address the **GOTO** goes to

(2) When using a debugger program, execution is stopped after which of the following debugger operations?
   (a) Step
   (b) User halt
   (c) Breakpoint
   (d) All of the above
   (e) None of the above

(3) Breakpoints are a good debugging tool for which of these programs?
   (a) Injection control for an automobile
   (b) Radio-controlled toy car
   (c) Water jets controller
   (d) Interactive handheld game
   (e) All of the above

(4) ICSP is used on a PIC® processor PCB for what?
   (a) Programming chips after they are soldered to the board
   (b) Debugging a program
   (c) Updating a program with a new version
   (d) All of the above
   (e) None of the above

(5) Rate the following debugging tools in order, from those that make the least impact on the program speed to those that make the most impact.

1. Breakpoints and single stepping
2. Profiling
3. Monitor

(a) 1, 2, 3
(b) 3, 2, 1
(c) 2, 3, 1
(d) 1, 3, 2
(e) 3, 1, 2

(6) Which of the following is not required to use data streaming?
(a) ICD
(b) Debugger
(c) ICSP connector
(d) PIC® processor
(e) PC

(7) An ICD unit is used for what with a PIC® processor?
(a) Counting and timing
(b) Programming and debugging
(c) Input and output
(d) Clocking and digitizing
(e) None of the above

(8) Which of the following would a profiling tool not be able to do?
(a) Find out how long functions take to execute
(b) Provide a method to change program calibration numbers
(c) Find out when and in what order functions are called
(d) Show the values of variables at key points in the program
(e) Verify all areas of code are executed

(9) Which of the following debugging tools would work with the E3 board without an ICD?
(a) Breakpoints and stepping
(b) Data streaming
(c) Monitor
(d) Profiling
(e) None of the above

(10) When an automobile mechanic plugs a tool into a port on a vehicle, what would be the best description of that port?
(a) Diagnostic interface
(b) Profiling port
(c) Debugging interface
(d) Monitor port
(e) ICSP

Bibliography

24AA01/24LC01B 1K I^2C Serial EEPROM Data Sheet. Microchip Technology.

25LC040 1K–4K SPI Serial EEPROM HighTemp Family Data Sheet. Microchip Technology.

American Standard Code for Information Interchange (ASCII), 1986. Std. ANSI-X3.4-1986(R1997). ANSI.

Electrical Characteristics of Balanced Voltage Digital Interface Circuits. TIA Std. TIA-422 (formally RS-422). Telecommunications Industry Association.

Electrical Characteristics of Generators and Receivers for Use in Balanced Digital Multipoint Systems. TIA Std. TIA-485 (formally RS-485). Telecommunications Industry Association.

IEEE Standard for Binary Floating-Point Arithmetic, 1985. IEEE Std 754–1985. IEEE Computer Society.

Interface Between Data Terminal Equipment and Data Circuit-Terminating Equipment Employing Serial Binary Data Interchange. TIA Std. TIA-232 (formally RS-232). Telecommunications Industry Association.

International Standard for Information Systems—Programming Language C. Std. ANSI/ISO 9899–1990, 1992. American National Standard for Information Systems.

International Standard for Information Systems—Programming Language C. Std. ANSI/ISO 9899–1999(E). American National Standard for Information Systems.

Kernighan, B.W., Ritchie, D.M., 1988. The C Programming Language, second ed. Prentice Hall, Englewood Cliffs, NJ.

Knuth, D., 1998. Art of Computer Programming, second ed., Sorting and Searching, vol. 3. Addison-Wesley, Reading, MA.

MAX220-MAX249, +5V-Powered, Multichannel RS-232 Drivers/Receivers Data Sheet. Maxim Integrated.

NMEA 0183 Manufacturer's Mnemonic Code. Std. National Marine Electronics Association.

PIC18F/LF1XK50 Data Sheet. Microchip Technology.

Plauger, P.J., 1991. The Standard C Library. Prentice Hall, Upper Saddle River, NJ.

Programming Languages – C – Extensions to Support Embedded Processors. 2008. Std. ISO/IEC TR 18037. International Organization for Standardization.

Schildt, H., 1993. The Annotated ANSI C Standard. Osborne McGraw-Hill, Berkeley, CA.

System Management Bus (SMBus) Control Method Interface Specification Version 1.0. System Management Interface Forum (SMIF) Inc.

TC74 Tiny Serial Digital Thermal Sensor Data Sheet. Microchip Technology.

The I^2C-Bus Specification. Philips Semiconductors.

The Unicode Standard – Version 4.0, defined by The Unicode Standard, Version 4.0, 2003. The Unicode Consortium. Addison-Wesley, Boston, MA.

UTF-8, A Transformation Format of ISO 10646, 2003. Std RFC 3629. The Internet Society.

Embedded C Programming. http://dx.doi.org/10.1016/B978-0-12-801314-4.00031-4

ASCII Chart

Dec	Hex	Sym	Dec	Hex	Sym	Dec	Hex	Sym	Dec	Hex	Sym	Dec	Hex	Dec	Hex	Dec	Hex	Dec	Hex
0	00	NUL	32	20	SP	64	40	@	96	60	`	128	80	160	A0	192	C0	224	E0
1	01	SOH	33	21	!	65	41	A	97	61	a	129	81	161	A1	193	C1	225	E1
2	02	STX	34	22	"	66	42	B	98	62	b	130	82	162	A2	194	C2	226	E2
3	03	ETX	35	23	#	67	43	C	99	63	c	131	83	163	A3	195	C3	227	E3
4	04	EOT	36	24	$	68	44	D	100	64	d	132	84	164	A4	196	C4	228	E4
5	05	ENQ	37	25	%	69	45	E	101	65	e	133	85	165	A5	197	C5	229	E5
6	06	ACK	38	26	&	70	46	F	102	66	f	134	86	166	A6	198	C6	230	E6
7	07	BEL	39	27	‘	71	47	G	103	67	g	135	87	167	A7	199	C7	231	E7
8	08	BS	40	28	(	72	48	H	104	68	h	136	88	168	A8	200	C8	232	E8
9	09	HT	41	29	)	73	49	I	105	69	i	137	89	169	A9	201	C9	233	E9
10	0A	LF	42	2A	*	74	4A	J	106	6A	j	138	8A	170	AA	202	CA	234	EA
11	0B	VT	43	2B	+	75	4B	K	107	6B	k	139	8B	171	AB	203	CB	235	EB
12	0C	FP	44	2C	'	76	4C	L	108	6C	l	140	8C	172	AC	204	CC	236	EC
13	0D	CR	45	2D	-	77	4D	M	109	6D	m	141	8D	173	AD	205	CD	237	ED
14	0E	SO	46	2E	.	78	4E	N	110	6E	n	142	8E	174	AE	206	CE	238	EE
15	0F	SI	47	2F	/	79	4F	O	111	6F	o	143	8F	175	AF	207	CF	239	EF
16	10	DLE	48	30	0	80	50	P	112	70	p	144	90	176	B0	208	D0	240	F0
17	11	DC1	49	31	1	81	51	Q	113	71	q	145	91	177	B1	209	D1	241	F1
18	12	DC2	50	32	2	82	52	R	114	72	r	146	92	178	B2	210	D2	242	F2
19	13	DC3	51	33	3	83	53	S	115	73	s	147	93	179	B3	211	D3	243	F3
20	14	DC4	52	34	4	84	54	T	116	74	t	148	94	180	B4	212	D4	244	F4
21	15	NAK	53	35	5	85	55	U	117	75	u	149	95	181	B5	213	D5	245	F5
22	16	SYN	54	36	6	86	56	V	118	76	v	150	96	182	B6	214	D6	246	F6
23	17	ETB	55	37	7	87	57	W	119	77	w	151	97	183	B7	215	D7	247	F7
24	18	CAN	56	38	8	88	58	X	120	78	x	152	98	184	B8	216	D8	248	F8
25	19	EM	57	39	9	89	59	Y	121	79	y	153	99	185	B9	217	D9	249	F9
26	1A	SUB	58	3A	:	90	5A	Z	122	7A	z	154	9A	186	BA	218	DA	250	FA
27	1B	ESC	59	3B	;	91	5B	[	123	7B	{	155	9B	187	BB	219	DB	251	FB
28	1C	FS	60	3C	<	92	5C	\	124	7C	|	156	9C	188	BC	220	DC	252	FC
29	1D	GS	61	3D	=	93	5D	]	125	7D	}	157	9D	189	BD	221	DD	253	FD
30	1E	RS	62	3E	>	94	5E	^	126	7E	~	158	9E	190	BE	222	DE	254	FE
31	1F	US	63	3F	?	95	5F	_	127	7F	DEL	159	9F	191	BF	223	DF	255	FF

Embedded C Programming. http://dx.doi.org/10.1016/B978-0-12-801314-4.00032-6

APPENDIX B

Statement	Example

if (expr) stmt; [else stmt;]	if (x= =25)
	  x=0;
	else
	  x=x+1

while (expr) stmt;	while (get_rtcc()!=0) putc('n');

do stmt while (expr);	do {
	  putc(c=getc());
	} while (c!=0);

for (expr1;expr2;expr3) stmt;	for (i-1;i<=10;++i) printf("%u/r/n",i);

switch (expr) {
case cexpr: stmt; //one or more case
[default:stmt]
...}
	switch (cmd) {
	  case 0: printf("cmd 0");break;
	  case 1: printf("cmd 1");break;
	  default: printf("bad cmd");break;
	}

return [expr];	return (5);
goto label;	goto loop;
label:stmt;	loop: i++;
break;	break;
continue;	continue;
expr;	i=1;
;	;
{[zero or more statements]}	{a=1; b=1;}
declaration;	int i;

Note: Items in [ ] are optional

Embedded C Programming. http://dx.doi.org/10.1016/B978-0-12-801314-4.00033-8

E3 Board Additional Information

The exercises in this book have been tailored to the CCS E3mini development board. This board uses the PIC18F14K50 processor. This board has a bootloader so no device programmer is required to reprogram the board with new software. It also has a USB port that can be used to communicate between a PC and the user program running on the PIC® (see Figure C.1).

CCS is providing an IDE and compiler to owners of this book at no additional cost that will work for the PIC18F14K50. The software can be downloaded using the following web link: www.ccsinfo.com/e3book.

Hardware

A schematic of the E3mini board is at the end of this appendix. The above web page also has information for purchasing the pre-built development board and a link to the PIC18F14K50 data sheet. Other development boards and even a simple bread-boarded PIC® can be used for these exercises as well. Pin designations and other instructions may need to be modified depending on the specific PIC® used and the development board configuration.

If building an E3mini-style board from scratch, a device programmer will be needed to load firmware into the part. Instructions are as follows:

1. Download the E3mini firmware image (.hex file) from the above web page.
2. Connect the device programmer to your target chip and the PC.
3. Power up the target board.
4. Start the device programmer software.
5. Load the e3mini.hex file.
6. Click on the "Write to Chip" button.

Software Install

After downloading the software simply execute the installer and follow the on-screen instructions to install. After installation there should be a desktop icon that looks like the one shown in Figure C.2 to start the IDE.

Embedded C Programming. http://dx.doi.org/10.1016/B978-0-12-801314-4.00034-X

Figure C.1: E3 prototyping board.

Figure C.2: CCS C compiler icon.

Compiling and Running a Program

1. Double-click on the compiler icon.
2. If a file opens up in the IDE, click on FILE>CLOSE ALL to clear the IDE.
3. Select FILE>NEW>SOURCE to start a new project. Select a name like EX0.C for the file. Notice the default project directory the compiler uses. Additional directories may be created under here. It is recommended to establish a unique file-naming convention or use a completely different directory for the source (see Figure C.3).
4. In the editor, type in the following program:

```
#include <e3.h>
#include <ios.h>

   void main(void) {
      cout << "Hello World !!!" << endl;
   }
```

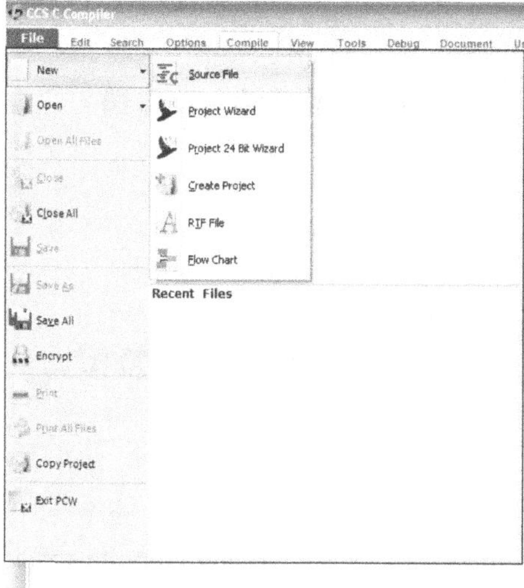

Figure C.3: CCS C compiler file menu.

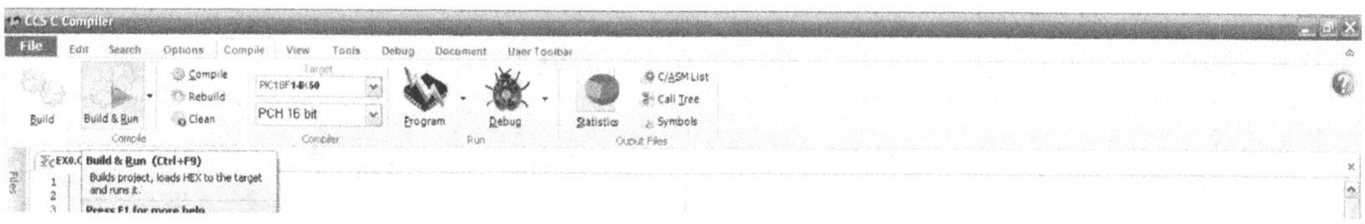

Figure C.4: CCS C compiler build and run on ribbon.

5. Connect the E3 board to the PC using a USB cable. If Windows indicates a new device has been found, click the default options on all the device wizard windows.
6. Click on BUILD AND RUN on the compile ribbon (see Figure C.4).
7. If the program is correct, after being compiled, the compile screen will show "No Errors" (see Figure C.5).
8. You will also see, in the lower right, your memory usage.
9. Next, a pop-up window will show the device being programmed.
10. Another pop-up window shows the output from the program. That window should show: Hello World!!! (see Figure C.6).

Figure C.5: CCS C compiler compile screen.

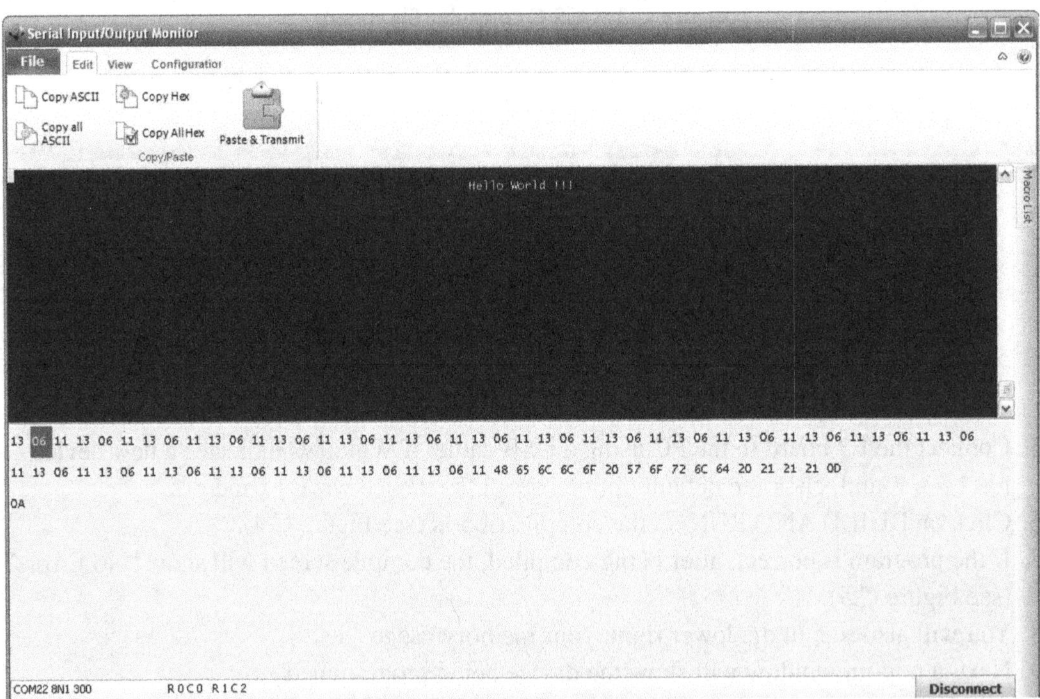

Figure C.6: CCS C compiler output window.

Note that for programs that do not have text I/O, close this pop-up window. The compiler always opens the window with the assumption the e3.h header will do text I/O.